T0235025

Multifunctional Cement-Based Materials

Civil and Environmental Engineering

A Series of Reference Books and Textbooks

Editor

Michael D. Meyer

Department of Civil and Environmental Engineering
Georgia Institute of Technology
Atlanta, Georgia

Multifunctional Cement-Based Materials

Deborah D. L. Chung

University at Buffalo
The State University of New York
Buffalo, New York, U.S.A.

CRC Press
Taylor & Francis Group
Boca Raton London New York

CRC Press is an imprint of the
Taylor & Francis Group, an **informa** business

CRC Press
Taylor & Francis Group
6000 Broken Sound Parkway NW, Suite 300
Boca Raton, FL 33487-2742

First issued in paperback 2019

© 2003 by Taylor & Francis Group, LLC
CRC Press is an imprint of Taylor & Francis Group, an Informa business

No claim to original U.S. Government works

ISBN-13: 978-0-8247-4610-0 (hbk)
ISBN-13: 978-0-367-39512-4 (pbk)

Visit the Taylor & Francis Web site at
http://www.taylorandfrancis.com

and the CRC Press Web site at
http://www.crcpress.com

In appreciation of my extended family

1st row: cousins of the author

2nd row: grand parents (three at the center of the row), great aunts, great uncles, and sister (baby) of the author

3rd row: parents (at the ends of the row), uncles and aunts of the author

4th row: uncles and cousin of the author

Photo taken in Hong Kong in 1950

Preface

Cement-based materials have been used for civil structures for many centuries. They remain dominant in today's structures. However, cement-based materials are now not only structural materials, but are also functional materials. The functions include sensing, which is relevant to smart structures, structural vibration control, traffic monitoring and structural health monitoring. They also include damping, thermal insulation, electrical grounding, resistance heating (say, for de-icing), controlled electrical conduction, electromagnetic interference shielding, lateral guidance of vehicles, electric current rectification (pn-junction), thermoelectricity, piezoresistivity, etc. These functions are rendered by the cement-based materials themselves, rather than embedding devices in cement-based materials. Materials science is used in attaining these functions. In contrast to existing books, which are focused on the structural properties of cement-based materials, this book is focused on the functional properties.

This book assumes that the reader has had only one introductory course on materials science. Thus, it covers the background concepts on cement-based materials and on non-structural behavior, in addition to providing the state of the art on functional cement-based materials. Numerous up-to-date references are provided in each chapter. The book is suitable for use as a textbook for senior undergraduate and graduate students in engineering (particularly materials, civil and mechanical engineering) and as a reference book for technicians, engineers and researchers.

Deborah D. L. Chung
Buffalo, NY
ddlchung@eng.buffalo.edu

Contents

1

Introduction to Cement-Based Materials

1.1   Background on cement-based materials

Composite materials are multi-phase materials obtained by artificial combination of different materials in order to attain properties that the individual components by themselves cannot attain. An example is a lightweight structural composite that is obtained by embedding continuous carbon fibers in one or more orientations in a polymer matrix. The fibers provide the strength and stiffness, while the polymer serves as the binder. Another example is concrete, which is a structural composite obtained by combining (through mixing) cement (the matrix, i.e., the binder, obtained by a reaction, known as hydration, between cement and water), sand (fine aggregate), gravel (coarse aggregate) and other optional ingredients that are known as admixtures. Short fibers and silica fume (a fine SiO_2 particulate) are examples of admixtures. In general, composite are classified according to their matrix material. The main classes of composites are polymer-matrix, cement-matrix, carbon-matrix and ceramic-matrix composites.

Cement-based materials include concrete (containing coarse and fine aggregates), mortar (containing fine aggregate but no coarse aggregate) and cement paste (containing no aggregate, whether coarse or fine). The fine aggregate is typically sand. The coarse aggregate is typically stones such as gravel. These aggregates are chosen due to their low cost and wide availability. The combination of coarse and fine aggregates allows dense packing of the aggregates as the fine aggregate fills the space between the units of large aggregate, as illustrated in Fig. 1.1. Both aggregates serve as fillers, while cement serves as the matrix (i.e., the binder). Hence, these materials are cement-matrix composites. Concrete is the form that is most commonly used in structures. Mortar is used in masonry (i.e., for joining bricks in a brick wall), coating and some forms of repair. Cement paste by itself is not used in structural applications, but is relevant to functional applications and is a basic component of concrete and mortar.

Cement is a silicate material, which microscopically involves interconnected tetrahedra with a silicon atom at the body center of each tetrehedron and an oxygen atom at each of the four corners of the tetrahedron (Fig. 1.2). In portland cement, the principal constituents are tricalcium silicate ($3CaO\text{-}SiO_2$) and dicalcium silicate ($2CaO\text{-}SiO_2$). When mixed with water, hydraulic cement such as portland cement forms a paste, which subsequently

1

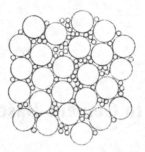

Fig. 1.1   Dense packing attained by the combined use of coarse and fine aggregates. Small circles: fine aggregate; large circles: coarse aggregate.

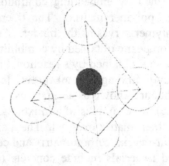

Fig. 1.2   A silicate tetrahedron. Solid circle: silicon atom; open circle: oxygen atom.

sets and hardens, thereby allowing cement to be used conveniently as the matrix in concrete structures of essentially any shape. The setting and hardening are due to hydration reactions between silicate and water. An example is the following:

$$2CaO - SiO_2 + xH_2O = 2CaO - SiO_2 - xH_2O,$$

where x depends on the amount of water available. The product of a hydration reaction is a hydrate in the form of a gel or a crystalline phase. The hydrate is responsible for a cementitious bond between cement and the aggregate. Setting refers to the stiffening of the plastic paste and takes several hours. It is followed by hardening, which continues for several years, although most of it occurs within a month. Although the material becomes dry as the water is used up in the hydration reaction, the apparent drying is not due to the loss of water by evaporation.

Due to the presence of ions such as Ca^{2+}, cement is alkaline. As a result, it can react slightly with the aggregate. This reaction is a cause of degradation of concrete.

Other than the aggregates, fillers in smaller quantities can be optionally added to the cement mix to improve the properties of the resulting materials. These fillers are called admixtures, which are discontinuous so that they can be included in the mix. They can be particles, such as silica fume (a fine particulate) and latex (a polymer in the form of a dispersion). They can be short fibers, such as polymer, steel, glass or carbon fibers. They can also be liquids such as methylcellulose aqueous solution, water reducing agent, defoamer, etc.

Cement-based materials also include steel-reinforced concrete, which is concrete containing steel-reinforcing bars (called rebars). The steel reinforcement is mainly used in the part of a concrete structure which encounters tension during use. This is because concrete is a brittle material and is much weaker under tension than compression. A rebar typically has surface deformations like ridges, so that mechanical interlocking occurs between the rebar and the concrete.

1.2   Improving cement-based materials by using silica fume

1.2.1   Introduction

Cement-based materials such as concrete have long been used for the civil infrastructure, such as highways, bridges and buildings. However, the deterioration of the civil infrastructure all over the U.S. has led to the realization that cement-based materials must be improved in terms of their properties and durability. The fabrication of cement-based materials involves mixing cement, water, aggregates (such as sand and gravel) and optionally other additives (called admixtures). The use of admixtures is a relatively convenient way of improving cement-based materials. Techniques involving special mixing, casting or curing procedures tend to be less attractive, due to the need for special equipment in the field. An admixture which is particularly effective is silica fume [1-4], although there are other admixtures such as latex and short fibers which are more expensive. This section is focused on the use of silica fume to improve cement-based materials.

Silica fume is very fine noncrystalline silica produced by electric arc furnaces as a by-product of the production of metallic silicon or ferrosilicon alloys. It is a powder with particles having diameters 100 times smaller than those of anhydrous portland cement particles, i.e., mean particle size between 0.1 and 0.2 μm. The SiO_2 content ranges from 85 to 98%. Silica fume is pozzolanic, i.e., it is reactive, like volcanic ash.

The property improvements needed for cement-based materials include increases in strength, modulus and ductility (whether under tension, compression, flexure or torsion), decrease in the drying shrinkage (i.e., shrinkage during the curing and simultaneous drying of a cement mix – a phenomenon which can cause cracks to form), decrease in the permeability to liquids and chloride ions (important for the corrosion resistance of the steel reinforcing bars embedded in concrete), and increase in the durability to freeze-thaw temperature cycling (important in cold regions).

Silica fume used as an admixture in a concrete mix has significant effects on the properties of the resulting material. These effects pertain to the strength, modulus, ductility, vibration damping capacity, sound absorption, abrasion resistance, air void content, shrinkage, bonding strength with reinforcing steel, permeability, chemical attack resistance, alkali-silica reactivity reduction, corrosion resistance of embedded steel reinforcement, freeze-thaw durability, creep rate, coefficient of thermal expansion (CTE), specific heat, thermal conductivity, defect dynamics, dielectric constant, and degree of fiber dispersion in mixes containing short microfibers. In addition, silica fume addition degrades the workability of the mix.

For the sake of consistency, the data given in this review to illustrate the effects of silica fume are all for silica fume from Elkem Materials Inc., Pittsburgh, PA (EMS 965), used in the amount of 15% by weight of cement. The cement is portland cement (Type I) from Lafarge Corp. (Southfield, MI). Curing is in air at room temperature and a relative humidity of 100% for 28 days. The water-reducing agent, if used, is a sodium salt of a condensed naphthalenesulphonic acid from Rohm and Haas Co., Philadelphia, PA (TAMOL SN).

1.2.2   Workability

Silica fume causes workability and consistency losses [5-18], which are barriers against proper utilization of silica fume concrete. However, the consistency of silica fume mortar is greatly enhanced by either using silane-treated silica fume, i.e., silica fume which has been coated by a silane coupling agent prior to incorporation in the mix, or using silane as an additional admixture [19,20]. The effectiveness of silane for cement is due to the reactivity of its molecular ends with –OH groups and the presence of –OH groups on the surface of both silica and cement.

Table 1.1 [19,20] shows that the silane introduction using either coating or admixture methods causes the silica fume mortar mix to increase in workability (slump). With silane (by either method) and no water-reducing agent, the workability of silica fume mortar mix is better than that of the mix with as-received silica fume and water-reducing agent in the amount of 1% by weight of cement. With silane (by either method) and a water-reducing agent in

Table 1.1 Workability of mortar mix[†]

Silica fume	Water-reducing agent/cement	Slump (mm)
Plain	0%	*
With untreated silica fume	0%	150
With untreated silica fume	1%	186
With untreated silica fume	2%	220
With treated silica fume	0%	194
With treated silica fume	0.2%	215
With silane and untreated silica fume	0%	197
With silane and untreated silica fume	0.2%	218

* Too large to be measured.
[†] The water/cement ratio was 0.35.

the amount of 0.2% by weight of cement, the workability is almost as good as the mix with as-received silica fume and water-reducing agent in the amount of 2% by weight of cement.

The increase in workability due to silane introduction is due to the improved wettability of silica fume by water. The improved wettability is expected from the hydrophylic nature of the silane molecule. Silane treatment involves formation of a silane coating on the surface of the silica fume; it does not cause surface roughening [21].

1.2.3 Mechanical properties

The most well-known effect of silica fume is the increase in strength [17,18,20,22-52], including the compressive strength [17,20,53-81], tensile strength [20,60,63,71,72,82] and flexural strength [59,66,72,83,84]. The strengthening is due to the pozzolanic activity of silica fume causing improved strength of the cement paste [36,79], the increased density of mortar or concrete resulting from the fineness of silica fume and the consequent efficient reaction to form hydration products, which fill the capillaries between cement and aggregate [50], the refined pore structure [55,76] and the microfiller effect of silica fume [76,77]. In addition, the modulus is increased [20,44,48,53,73,85]. These effects are also partly due to the densification of the interfacial zone between paste and aggregate [85].

As shown in Table 1.2 [20] for cement pastes at 28 days of curing, the tensile strength, tensile ductility, compressive strength and compressive modulus are increased and the compressive ductility is decreased by the addition of silica fume (15% by weight of cement) which has not been surface treated. The tensile strength and compressive strength are further increased and the compressive ductility is further decreased when silane-treated silica fume is

Table 1.2 Mechanical properties of cement pastes at 28 days of curing

	Plain	With untreated silica fume[†]	With treated silica fume[†]	With silane[*] and untreated silica fume[†]
Tensile strength (MPa)	0.91 ± 0.02	1.53 ± 0.06	2.04 ± 0.06	2.07 ± 0.05
Tensile modulus (GPa)	11.2 ± 0.24	10.2 ± 0.7	11.5 ± 0.6	10.9 ± 0.5
Tensile ductility (%)	0.0041 ± 0.00008	0.020 ± 0.0004	0.020 ± 0.0004	0.021 ± 0.0004
Compressive strength (MPa)	57.9 ± 1.8	65.0 ± 2.6	77.3 ± 4.1	77.4 ± 3.7
Compressive modulus (GPa)	2.92 ± 0.07	13.6 ± 1.4	10.9 ± 1.8	15.8 ± 1.6
Compressive ductility (%)	1.72 ± 0.04	0.614 ± 0.023	0.503 ± 0.021	0.474 ± 0.015

[*] 0.2% by weight of cement.
[†] 15% by weight of cement.

used. On the other hand, the tensile modulus is essentially not affected by the silica fume addition. The use of both silane (0.2% by weight of cement) and untreated silica fume (as two admixtures, last column of Table 1.2) enhances the tensile strength, compressive strength and compressive modulus, but decreases the compressive ductility, relative to the paste with untreated silica fume and no silane. The effect of using treated silica fume (next to the last column of Table 1.2) and of using the combination of silane and untreated silica fume are quite similar, except that the compressive modulus is higher and the compressive ductility is lower for the latter due to the network of covalent silane coupling among the silica fume particles in the latter case.

The use of silane as an admixture which is added directly into the cement mix involves slightly more silane material but less processing cost than the use of silane in the form of a coating on silica fume. Both methods of silane introduction result in increases in the tensile and compressive strengths. The network attained by the admixture method of silane introduction does not result from the silane coating method, due to the localization of the silane in the coating, which nevertheless provides chemical coupling between silica fume and cement. The network, which is formed from the hydrolysis and polymerization (condensation) reaction of silane during the hydration of cement, also causes the ductility to decrease.

1.2.4 Vibration damping capacity

Vibration reduction is valuable for hazard mitigation, structural stability and structural performance improvement. Effective vibration reduction requires both damping capacity and stiffness. Silica fume is effective for enhancing both damping capacity and stiffness (modulus) [20,86-88].

As shown in Table 1.3 [20], the vibration damping capacity, as expressed by the loss tangent under dynamic 3-point flexural loading at 0.2 Hz, is significantly increased by the addition of silica fume which has not been surface treated. The use of silane-treated silica fume increases the loss tangent slightly beyond the value attained with untreated silica fume. The use of silane and untreated silica fume as two admixtures decreases the loss tangent to a value below that attained by using untreated silica fume alone, but still above that for plain cement paste.

The ability of silica fume to enhance the damping capacity is due to the large area of the interface between silica fume particles and the cement matrix and the contribution of interface slippage to energy dissipation. Although the pozzolanic nature of silica fume makes the interface rather diffuse, the interface still contributes to damping. The silane covalent coupling introduced by the silane surface treatment of silica fume can move during vibration, thus providing another mechanism for damping and enhancing the loss tangent. The network introduced by the use of silane and untreated silica fume as two admixtures restricts movement and therefore reduces the damping capacity relative to the case with untreated silica fume alone. Nevertheless, the use of the two admixtures enhances the damping capacity relative to the plain cement paste case, as even less movement is possible in plain cement paste.

The storage modulus (Table 1.3) is significantly increased by the addition of untreated silica fume, is further increased by the use of silane treated silica fume, and is still further increased by the use of silane and untreated silica fume as two admixtures. The increase in storage modulus upon addition of untreated silica fume is attributed to the high modulus of silica compared to the cement matrix. The enhancement of the storage modulus by the use of silane-treated silica fume is due to the chemical coupling provided by the silane between silica fume and cement. The further enhancement of the storage modulus by the use of silane and untreated silica fume as two admixtures is due to the network of covalent coupling among the silica fume particles.

The loss modulus (Table 1.3) is the product of the loss tangent and the storage modulus. As vibration reduction requires both damping and stiffness, both the loss tangent and storage modulus should be high for effective vibration reduction. Hence, the loss modulus serves as an overall figure of merit for the vibration reduction ability. The loss modulus is much increased by the addition of untreated silica fume and is further increased by the use of silane treated silica fume. However, the use of silane and untreated silica fume as two admixtures

Table 1.3   Dynamic flexural properties of cement pastes at a flexural (3-point bending) loading frequency of 0.2 Hz

	Loss tangent (tan δ, ± 0.002)	Storage modulus (GPa, ± 0.03)	Loss modulus (GPa, ± 0.02)
Plain	0.035	1.91	0.067
With untreated silica fume[†]	0.082	12.71	1.04
With treated silica fume[†]	0.087	16.75	1.46
With silane[*] and untreated silica fume[†]	0.055	17.92	0.99

[*] 0.2% by weight of cement.
[†] 15% by weight of cement.

decreases the loss modulus to a value below the paste with untreated silica fume alone, due to the decrease in the loss tangent. As a result, the use of silane-treated silica fume gives the highest value of the loss modulus.

1.2.5   Sound absorption

Sound or noise absorption is useful for numerous structures, such as pavement overlays and noise barriers. The addition of silica fume to concrete improves the sound absorption ability [88]. The effect is related to the increase in vibration damping capacity (previous section).

1.2.6   Freeze-thaw durability

Freeze-thaw durability refers to the ability to withstand changes between temperatures above 0°C and those below 0°C. Due to the presence of water, which undergoes freezing and thawing (which in turn cause changes in volume), concrete tends to degrade upon such temperature cycling. Air voids (called air entrainment) are used as cushions to accommodate the changes in volume, thereby enhancing the freeze-thaw durability.

The addition of silica fume to mortar improves the freeze-thaw durability [89-92], in spite of the poor air-void system [92]. However, the use of air entrainment is still recommended [93-97]. The addition of silica fume also reduces scaling [98].

1.2.7   Abrasion resistance

The addition of untreated silica fume to mortar increases the abrasion (both solid and hydraulic) resistance [48,98,99], as shown by the depth of wear decreasing from 1.07 to 0.145 mm (as tested using ASTM C944-90a, rotating-cutter method) [99]. The abrasion resistance is further improved by using acid treated silica fume [100].

1.2.8 Shrinkage

The hydration reaction that occurs during the curing of cement causes shrinkage, called autogenous shrinkage, which is accompanied by a decrease in the relative humidity within the pores. When the curing is conducted in an open atmosphere, as is usually the case, additional shrinkage occurs due to the movement of water through the pores to the surface and the loss of water on the surface by evaporation. The overall shrinkage that occurs in this case is known as the drying shrinkage, which is the shrinkage that is important for practical purposes.

Drying shrinkage can cause cracking and prestressing loss [101]. The addition of untreated silica fume to cement paste decreases the drying shrinkage [20,34,101-105] (Table 1.4). This desirable effect is partly due to the reduction of the pore size and connectivity of the voids and partly due to the prestressing effect of silica fume, which restrains the shrinkage. The use of silane-treated silica fume in place of untreated silica fume further decreases the drying shrinkage, due to the hydrophylic character of the silane-treated silica fume and the formation of chemical bonds between silica fume particles and cement [20]. The use of silane and untreated silica fume as two admixtures also decreases the drying shrinkage, but not as significantly as the use of silane-treated silica fume [20]. However, silica fume has also been reported to increase the drying shrinkage [34,106,107] and the restrained shrinkage crack width is increased by silica fume addition [108].

Due to the pozzolanic nature of silica fume, silica fume addition increases the autogenous shrinkage, as well as the autogenous relative humidity change [109,110]. These effects are undesirable, as they may cause cracking if the deformation is restrained.

Carbonization refers to the chemical reaction between CO_2 and cement, as made possible by the in-diffusion of CO_2 gas. This reaction causes shrinkage,

Table 1.4 Drying shrinkage strain (10^{-4}, ± 0.015) of cement pastes at 28 days

Plain	4.98
With untreated silica fume	4.41
With treated silica fume	4.18
With silane and untreated silica fume	4.32

which is called carbonization shrinkage. Due to the effect of silica fume addition on the pore structure, which affects the in-diffusion, the carbonization shrinkage may be avoided by the addition of silica fume [111].

Concrete exposed to hot climatic conditions soon after casting is particularly prone to plastic shrinkage cracking [112], which is primarily due to the development of tensile capillary pressure during drying. Silica fume addition increases the plastic shrinkage [113], due to the high tensile capillary pressure resulting from the high surface area of the silica fume particles.

1.2.9   Air void content and density

The air void content of cement paste (Table 1.5) is increased by the addition of untreated silica fume [20,114]. Along with this effect is a decrease in density (Table 1.6). Both effects are related to the reduction in drying shrinkage. The introduction of silane by either the coating or admixture method decreases the air void content, but the value is still higher than that of plain cement paste. The use of the admixture method of silane introduction increases the density to a value almost as high as that of plain cement paste, due to the network of covalent coupling among the silica fume particles [20]. On the other hand, the air void content of concrete is decreased by silica fume addition [8,17], probably due to the densification of the paste-aggregate interface [85].

1.2.10   Permeability

The permeability of chloride ions in concrete is decreased by the addition of untreated silica fume [5,8,17,22,25,30-32,46,47,50,57,59,78,103,

Table 1.5   Air void content (%, ± 0.02) of cement pastes

Plain	2.32
With untreated silica fume	3.73
With treated silica fume	3.26
With silane and untreated silica fume	3.19

Table 1.6   Density (g/cm³, ± 0.02) of cement pastes

Plain	2.01
With untreated silica fume	1.72
With treated silica fume	1.73
With silane and untreated silica fume	1.97

115-138]. Related to this effect is the decrease in the water absorptivity. Both effects are due to the microscopic pore structure resulting from the calcium silicate hydrate (CSH) formed upon the pozzolanic reaction of silica fume with free lime during the hydration of concrete [139-163].

1.2.11 Steel rebar corrosion resistance

The addition of untreated silica fume to steel-reinforced concrete enhances the corrosion resistance of the reinforcing steel [163-177]. This is related to the decrease in the permeability (last section).

1.2.12 Alkali-silica reactivity reduction

The alkali-silica reactivity refers to the reactivity of silica (present in most aggregates) and alkaline ions (present in cement). It is detrimental due to the expansion caused by the reaction product. This reactivity is reduced by the addition of silica fume [178-195], because of the effectiveness of silica fume to remove alkali from the pore solution [184,185,190], to reduce the alkali ion (Na^+, K^+, OH^-) concentrations in the pore solution [193,195], and to retard the transportation of alkalis to reaction sites (due to the refinement and segmentation of the pore structure) [192]. However, silica fume with coarse particles or undispersed agglomerates can induce distress related to alkali-silica reactivity [196].

1.2.13 Chemical attack resistance

The addition of untreated silica fume to concrete enhances the chemical attack resistance [197-213], whether the chemical is acid, sulfate, chloride, etc. This effect is related to the decrease in permeability.

1.2.14 Bond strength to steel rebar

The addition of untreated silica fume to concrete increases the shear bond strength between concrete and steel rebar [214-220]. This effect is mainly due to the reduced porosity and thickness of the transition zone adjacent to the steel, thereby improving the adhesion-type bond at small slip levels [217-220]. The combined use of silica fume and methylcellulose (0.4% by weight of cement) gives even higher bond strength, due to the surfactant role of methylcellulose [218,221].

1.2.15 Creep rate

The addition of untreated silica fume to cement paste decreases the compressive creep rate at 200°C from 1.3×10^{-5} to 2.4×10^{-6} min^{-1} [114]. The creep resistance is consistent with the high storage modulus (Table 1.3), which remains much higher than that of plain cement paste up to at least 150°C [114]. However, silica fume increases the early age tensile creep, which provides a mechanism to relieve some of the restraining stress that develops due to autogenous shrinkage [222].

1.2.16 Coefficient of thermal expansion

The CTE is reduced by the addition of untreated silica fume [114]. This is consistent with the high modulus and creep resistance.

1.2.17 Specific heat

A high value of the specific heat is valuable for improving the temperature stability of a structure and to retain heat in a building. The specific heat (C_p, Table 1.7) is increased by the addition of untreated silica fume [114]. The use of silane-treated silica fume in place of untreated silica fume further increases the specific heat, though only slightly [223]. The effect of untreated silica fume is due to the slippage at the interface between the silica fume and cement. The effect of the silane treatment is due to the contribution of the movement of the covalent coupling between silica fume particles and the cement. The use of silane and untreated silica fume as two admixtures greatly increases the specific heat, due to the network of covalent coupling among the silica fume particles contributing to phonons [20].

1.2.18 Thermal conductivity

Concrete of low thermal conductivity is useful for the thermal insulation of buildings. On the other hand, concrete of high thermal conductivity is useful for reducing temperature gradients in structures. The thermal stresses that result from temperature gradients may cause mechanical

Table 1.7   Specific heat (J/g.K, ± 0.001) of cement pastes

Plain	0.736
With untreated silica fume	0.782
With treated silica fume	0.788
With silane and untreated silica fume	0.980

Table 1.8   Thermal conductivity (W/m.K, ± 0.07) of cement pastes

Plain	0.53
With untreated silica fume	0.35
With treated silica fume	0.33
With silane and untreated silica fume	0.61

property degradation and even warpage in the structure. Bridges are among structures that tend to encounter temperature differentials between their top and bottom surfaces. In contrast to buildings, which also encounter temperature differentials, bridges do not need thermal insulation. Therefore, concrete of high thermal conductivity is desirable for bridges and related structures.

The thermal conductivity (Table 1.8) is decreased by the addition of untreated or silane-treated silica fume [20,114,223], due to the interface between silica fume particles and cement acting as a barrier against heat conduction. However, the thermal conductivity is increased by the use of silane and untreated silica fume as two admixtures [20], due to the network of covalent coupling enhancing heat conduction through phonons.

1.2.19 Fiber dispersion

Short microfibers, such as carbon, glass, polypropylene, steel and other fibers, are used as an admixture in concrete to enhance the tensile and flexural properties and decrease the drying shrinkage. Effective use of the fibers, which are used in very small quantities (such as 0.5% by weight of cement in the case of carbon fibers), requires good dispersion of the fibers. The addition of untreated silica fume to microfiber reinforced cement increases the degree of fiber dispersion, due to the fine silica fume particles helping the mixing of the microfibers [224-269]. In addition, silica fume improves the structure of the fiber-matrix interface, reduces the weakness of the interfacial zone and decreases the number and size of cracks [269].

Table 1.9 shows that silica fume (without treatment) decreases the resistivity of carbon fiber mortar and increases the tensile strength, modulus and ductility [270]. The lower resistivity indicates better fiber dispersion.

1.2.20 Defect dynamics during elastic deformation

Defects in a solid respond to applied stresses. When the applied stress is dynamic, the response of the defects is also dynamic. The response encompasses the generation, healing and aggravation of defects. Defect generation refers to the formation of defects which usually occurs during loading. Defect healing refers to the diminution of defects. Healing can occur

Table 1.9   Tensile properties and electrical resistivity of mortars that contain carbon fibers and methylcellulose (0.4% by mass of cement). The water/cement ratio is 0.350.

	Without silica fume	With silica fume*
Electrical resistivity (10^6 Ω.cm)	0.68 (± 2.8%)	0.31 (± 3.1%)
Tensile strength (MPa)	2.26 ± 0.08	2.36 ± 0.06
Tensile modulus (GPa)	12.5 ± 0.3	13.5 ± 0.9
Tensile ductility (%)	0.0150 ± 0.0009	0.0168 ± 0.0009

* Without treatment.

during compressive loading of a brittle materials, such as a cement-based material.   Defect aggravation refers to the propagation or enlargement of defects; it can occur during removal of a compressive stress from a brittle material.

Figs. 1.3 and 1.4 show the variation of the fractional change in longitudinal resistivity with cycle no. during initial cyclic compression of plain mortar and silica fume (without treatment) mortar respectively in the elastic regime [271]. The stress amplitude used increased cycle by cycle. For both mortars, the resistivity increased abruptly during the first loading (due to defect generation) and increased further during the first unloading (due to defect aggravation).   Moreover, the resistivity decreased during subsequent loading (due to defect healing) and increased during subsequent unloading (due to defect aggravation); the effect associated with defect healing was much larger for silica fume mortar than for plain mortar. In addition, this effect intensified as stress cycling at increasing stress amplitudes progressed for both mortars, probably due to the increase in the extent of minor damage. The increase in damage extent was also indicated by the resistivity baseline increasing gradually cycle by cycle. In spite of the increase in stress amplitude cycle by cycle, defect healing dominated over defect generation during loading in all cycles other than the first cycle.

Comparison of Figs. 1.3 and 1.4 showed that silica fume contributed significantly to the defect dynamics during elastic deformation. The associated defects were presumably at the interface between silica fume and cement, even though this interface was diffuse due to the pozzolanic nature of silica fume. The defects at this interface were smaller than those at the sand-cement interface, but this interface was large in total area due to the small particle size of silica fume compared to sand.

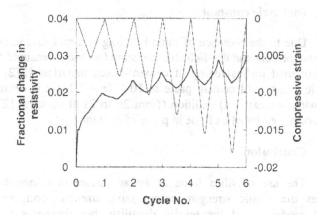

Fig. 1.3   Variation of the fractional change in resistivity with cycle no. (thick curve) and of the compressive strain with cycle no. (thin curve) during repeated compressive loading at increasing stress amplitudes within the elastic regime for plain mortar (without silica fume).

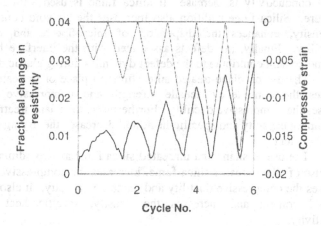

Fig. 1.4   Variation of the fractional change in resistivity with cycle no. (thick curve) and of the compressive strain with cycle no. (thin curve) during repeated compressive loading at increasing stress amplitudes within the elastic regime for silica fume mortar.

1.2.21   Dielectric constant

Due to the presence of ionic bonding and moisture in cement, electric dipoles are present and the dielectric constant has been measured for the purpose of fundamental understanding of cement-based materials [272]. The relative dielectric constant of cement paste at 10 kHz – 1 MHz is decreased by silica fume (without treatment) addition (from 29 to 21 at 10 kHz) [273], due to the volume occupied by silica fume in place of cement.

1.2.22   Conclusion

The use of silica fume as an admixture in cement-based materials increases the tensile strength, compressive strength, compressive modulus, flexural modulus and the tensile ductility, but decreases the compressive ductility. In addition, it enhances the freeze-thaw durability, the vibration damping capacity, the abrasion resistance, the bond strength with steel rebars, the chemical attack resistance and the corrosion resistance of reinforcing steel. It also decreases the alkali-silica reactivity, the drying shrinkage, permeability, creep rate, coefficient of thermal expansion and dielectric constant. Moreover, it increases the specific heat and decreases the thermal conductivity, though the thermal conductivity is increased if silica fume is used with silane, another admixture. Silica fume addition also increases the air void content, decreases the density, enhances the dispersion of microfibers, and decreases the workability. Finally, the defects associated with the interface between silica fume and cement contribute to the defect dynamics during elastic deformation.

The use of silane-treated silica fume in place of untreated silica fume increases the consistency, tensile strength and compressive strength, but decreases the compressive ductility. Furthermore, the silane treatment increases the damping capacity and specific heat, and decreases the drying shrinkage and air void content.

The use of silane and untreated silica fume as two admixtures, relative to the use of silane-treated silica fume, increases the compressive modulus, but decreases the compressive ductility and damping capacity. It also decreases the air void content and increases the density, specific heat and thermal conductivity.

1.3   Improving cement-based materials by using short fibers

The admixtures in a cement-based material can be a fine particulate such as silica (SiO_2) fume for decreasing the porosity in the composite. It can be a polymer (used in either a liquid solution form or a solid dispersion form) such as latex, also for decreasing the porosity. It can be short fibers (such as carbon fibers, glass fibers, polymer fibers and steel fibers) for increasing the toughness and decreasing the drying shrinkage (shrinkage during curing – undesirable as it can cause cracks to form). Continuous fibers are seldom used because of their high cost and the impossibility of incorporating continuous fibers in a cement mix. Due to the bidding system for many construction projects, low cost is essential.

Carbon fiber cement-matrix composites are structural materials that are quite rapidly gaining in importance due to the decrease in carbon fiber cost [274] and the increasing demand for superior structural and functional properties. These composites contain short carbon fibers, typically 5 mm in length. However, due to the weak bond between carbon fiber and the cement matrix, continuous fibers [275-277] are much more effective than short fibers in reinforcing concrete. Surface treatment of carbon fiber (e.g., by heating [278] or by using ozone [279,280], silane [281], SiO_2 particles [282] or hot NaOH solution [283]) is useful for improving the bond between fiber and matrix, thereby improving the properties of the composite. In the case of surface treatment by ozone or silane, the improved bond is due to the enhanced wettability by water. Admixtures such as latex [279,284], methylcellulose [279] and silica fume [285] also help the bond.

The effect of carbon fiber addition on the properties of concrete increases with fiber volume fraction [286], unless the fiber volume fraction is so high that the air void content becomes excessively high [287]. (The air void content increases with fiber content and air voids tend to have a negative effect on many properties, such as the compressive strength.) In addition, the workability of the mix decreases with fiber content [286]. Moreover, the cost increases with fiber content. Therefore, a rather low volume fraction of fibers is desirable. A fiber content as low as 0.2 vol. % is effective [288], although fiber contents exceeding 1 vol. % are more common [289,293]. The required fiber content increases with the particle size of the aggregate, as the flexural strength decreases with increasing particle size [294].

Effective use of the carbon fibers in concrete requires dispersion of the fibers in the mix. The dispersion is enhanced by using silica fume (a fine particulate) as an admixture [287,295-297]. A typical silica fume content is 15% by weight of cement [287]. The silica fume is typically used along with a small amount (0.4% by weight of cement) of methylcellulose for helping the dispersion of the fibers and the workability of the mix [287]. Latex (typically 15-20% by weight of cement) is much less effective than silica fume for helping the fiber dispersion, but it enhances the workability, flexural strength, flexural toughness, impact resistance, frost resistance and acid resistance [287,298,299]. The ease of dispersion increases with decreasing fiber length [287].

The improved structural properties rendered by carbon fiber addition pertain to the increased tensile and flexible strengths, the increased tensile ductility and flexural toughness, the enhanced impact resistance, the reduced drying shrinkage and the improved freeze-thaw durability [286-288,290-

298,300-311]. The tensile and flexural strengths decrease with increasing specimen size, such that the size effect becomes larger as the fiber length increases [312]. The low drying shrinkage is valuable for large structures and for use in repair [313,314] and in joining bricks in a brick structure [315,316]. The functional properties rendered by carbon fiber addition pertain to the strain sensing ability [280, 317-330] (for smart structures), the temperature sensing ability [331-334], the damage sensing ability [317, 321, 335-337], the thermoelectric behavior [332-334], the thermal insulation ability [338-340] (to save energy for buildings), the electrical conduction ability [341-350] (to facilitate cathodic protection of embedded steel and to provide electrical grounding or connection), and the radio wave reflection/absorption ability [351-355] (for electromagnetic interference or EMI shielding, for lateral guidance in automatic highways, and for television image transmission).

In relation to the structural properties, carbon fibers compete with glass, polymer and steel fibers [291, 300-302, 287, 309-311, 356]. Carbon fibers (isotropic pitch based) [274, 356] are advantageous in their superior ability to increase the tensile strength of concrete, even though the tensile strength, modulus and ductility of the isotropic pitch based carbon fibers are low compared to most other fibers. Carbon fibers are also advantageous in the relative chemical inertness [357]. PAN-based carbon fibers are also used [290, 292, 296, 307], although they are more commonly used as continuous fibers than short fibers. Carbon-coated glass fibers [358, 359] and submicron diameter carbon filaments [349-351] are even less commonly used, although the former is attractive for the low cost of glass fibers and the latter is attractive for its high radio wave reflectivity (which results from the skin effect). C-shaped carbon fibers are more effective for strengthening than round carbon fibers [360], but their relatively large diameter makes them less attractive. Carbon fibers can be used in concrete together with steel fibers, as the addition of short carbon fibers to steel fiber reinforced mortar increases the fracture toughness of the interfacial zone between steel fiber and the cement matrix [361]. Carbon fibers can also be used in concrete together with steel bars [362,363], or together with carbon fiber reinforced polymer rods [364].

In relation to most functional properties, carbon fibers are exceptional compared to the other fiber types. Carbon fibers are electrically conductive, in contrast to glass and polymer fibers, which are not conductive. Steel fibers are conductive, but their typical diameter (≥ 60 μm) is much larger than the diameter of a typical carbon fiber (15 μm). The combination of electrical conductivity and small diameter makes carbon fibers superior to the other fiber types in the area of strain sensing and electrical conduction. However, carbon fibers are inferior to steel fibers for providing thermoelectric composites, due to the high electron concentration in steel and the low hole concentration in carbon.

Although carbon fibers are thermally conductive, the addition of carbon fibers to concrete lowers the thermal conductivity [338], thus allowing applications related to thermal insulation. This effect of carbon fiber addition is due to the increase in air void content. The electrical conductivity of carbon fibers is higher than that of the cement matrix by about 8 orders of magnitude, whereas the thermal conductivity of carbon fibers is higher than that of the cement matrix by only one or two orders of magnitude. As a result, the electrical conductivity is increased upon carbon fiber addition in spite of the increase in air void content, but the thermal conductivity is decreased upon fiber addition.

1.4   Improving cement-based materials by interface engineering

1.4.1   Introduction

Interfaces are present in a composite material and greatly affect the properties of the composite material. Interface engineering refers to the design and preparation of an interface. This is widely practiced for composites with polymers, metals, ceramics and carbons as matrices in order to improve the properties of the composites. However, partly due to the importance of low cost for practical concretes, interface engineering has received relatively little attention in relation to cement-matrix composites. Nevertheless, the need for improved concretes in today's infrastructure is recognized by countries all over the world, thus resulting in a new momentum in research to improve cement-matrix composites. Interface engineering is a significant aspect of recent research on the improvement of cement-matrix composites. This engineering involves surface treatments of steel reinforcing bars (rebars) and of admixtures, as well as the use of admixtures. The surface treatment of aggregates is not economically practical, due to the low cost and large volume of usage of aggregates in concretes. Interface engineering for cement-matrix composites is covered in this section, with emphasis on the methods and effects of the interface engineering, rather than the mechanisms, due to the infancy of the field.

1.4.2   Steel rebar surface treatments for interface engineering

The interface between steel rebar and concrete affects the bond strength between rebar and concrete, in addition to affecting the corrosion resistance of the rebar in the concrete and affecting the vibration reduction ability of the steel-reinforced concrete. The shaping of the rebar (say, having surface deformations called ribs, or having a hook at an end of the rebar) enhances the mechanical interlocking between rebar and concrete, thereby improving the interface. However, this section focuses on rebar surface treatments rather than the shaping of the rebar for the purpose of interface engineering. The treatments include sandblasting [339-341], water immersion [340], ozone treatment [339-

341] and acetone washing [339]. Some of them are useful for improving the bond strength, vibration reduction ability and corrosion resistance, as described below.

Sandblasting involves the blasting of ceramic particles (typically alumina particles of size around 250 μm) under pressure (typically around 80 psi or 0.6 MPa). It results in roughening as well as cleaning of the surface of the steel rebar. The cleaning relates to the removal of rust and other contaminants on the rebar surface, as a steel rebar typically is covered by rust and other contaminants here and there. The cleaning causes the surface of the rebar to be more uniform in composition.

Water immersion means total immersion of the rebar in water at room temperature for two days. It causes the formation of a black oxide layer on the surface of the rebar. Water immersion times that are less than or greater than two days give less desirable effects on both bond strength and corrosion resistance.

Ozone treatment involves exposure of the rebar to ozone (O_3) gas (say, 0.3 vol. % in air) for 20 min at 160°C, followed by drying at 110°C in air for 50 min. It causes the formation of a dark gray oxide layer on the surface of the rebar.

Acetone washing involves immersion of rebar in acetone for 15 min, followed by drying in air. It removes the grease on the rebar surface.

Figures 1.5-1.7 show the correlation of the contact resistivity of the rebar-concrete interface with the shear bond strength for different surface treatments of steel rebar. The contact resistivity increases roughly linearly with increasing bond strength, such that the data for the different surface treatments lie on essentially parallel straight lines. Acetone treatment increases the bond strength slightly and decreases the contact resistivity slightly (Fig. 1.5) (as in the case of the interface between stainless steel fiber and cement paste [342]), presumably because of the degreasing action of the acetone. Water immersion for 2-5 days (Fig. 1.6) increases the bond strength by 14% (more than for acetone treatment) and slightly increases the contact resistivity (in contrast to the decrease in contact resistivity for acetone treatment). Increase of the water immersion time beyond 5 days causes the bond strength to decrease and the contact resistivity to increase further (Fig. 1.6). However, even for a water immersion time of 10 days, the bond strength is still higher than that for the as-received rebar. Thus, a water immersion time of 2 days is recommended. Fig. 1.6 shows that ozone treatment enhances the bond strength more than any of the water treatments. The contact resistivity is also increased by the ozone treatment, but not as much as in the case of water treatment for 7 or 10 days.

It is reasonable to assume that the contact resistivity is related to the amount of oxidation product at the rebar-concrete interface, as the oxidation product is a poor electrical conductor. Hence, the differences in contact resistivity (Fig. 1.6) suggest that the amount of oxidation product is comparable

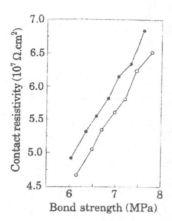

Fig. 1.5   Variation of contact electrical resistivity with bond strength between steel rebar and concrete at 28 days of curing. Solid circles: as-received steel rebar. Open circles: acetone-treated steel rebar.

between O_3 treatment and 2-5 day water treatments, but is larger for 7-10 day water treatments. The phase of the oxidation product differs between O_3 and water treatments, as indicated by the black color of the oxidation product of the water treatments and the dark gray color of the oxidation product of the O_3 treatment. This phase difference is believed to be partly responsible for the difference in the extent of bond strength enhancement.

The contact resistivity increases with increasing bond strength among the data for each water immersion time (Fig. 1.6). The origin of this dependence is associated with interfacial phase(s) of volume resistivity higher than that of concrete. The interfacial phase enhances the bonding, unless it is excessive. It may be a metal oxide. Water treatment increases both bond strength and contact resistivity because the treatment forms a black phase that may be akin to rust on the rebar; the phase enhances the bonding but increases the contact resistivity. The longer the water immersion time, the more the black phase and the higher the contact resistivity. However, an excessive amount of the black phase (as obtained after 7 or 10 days of water immersion) weakens the bond.

At the same bond strength, the water-treated rebar exhibits a lower contact resistivity than the as-received rebar (Fig. 1.6). As the amount of black phase increases with increasing contact resistivity, this implies that the black phase formed by the water treatment is more effective than the rust or rust-like phase(s) formed without the water treatment in enhancing the bond strength. The greater effectiveness of the former is probably partly because of the more uniform distribution of the black phase and partly because of the possible

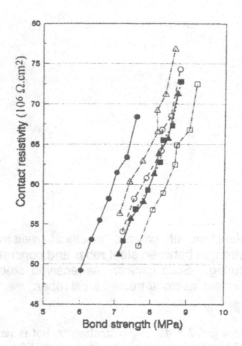

Fig. 1.6   Variation of contact electrical resistivity with bond strength between steel rebar and concrete at 28 days of curing. Solid circles: as-received steel rebar. Solid triangles: steel rebar immersed in water for 2 days. Solid squares: steel rebar immersed in water for 5 days. Open circles: steel rebar immersed in water for 7 days. Open triangles: steel rebar immersed in water for 10 days. Open squares: O_3-treated steel rebar.

differences in phase between the black phase and the rust or rust-like phase formed without the water treatment.

Water treatment and sandblasting increase the bond strength to similar extents (Fig. 1.7), which are less than that provided by ozone treatment (Fig. 1.6). Water immersion, like ozone treatment, causes the contact resistivity to increase, but sandblasting has negligible effect on the contact resistivity. This is consistent with the presence of a black coating on the rebar after water immersion and the absence of a coating after sandblasting. Scanning electron microscopy (SEM) shows that sandblasting roughens the surface in a coarse way, whereas water treatment results in a fine surface microstructure. The uneven surface quality (due to uneven rusting) in the as-received rebar is

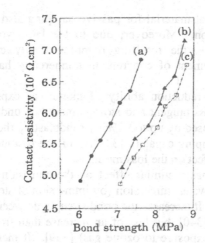

Fig. 1.7   Variation of contact electrical resistivity with bond strength between steel rebar and concrete at 28 days of curing. (a) As-received rebar. (b) Water-treated rebar. (c) Sandblasted rebar.

removed after sandblasting or water treatment, as shown by visual observation. In spite of the significant roughening by sandblasting, the bond strength is similar for the sandblasted rebar and the water treated rebar. This suggests that the bond strength increase after water immersion is essentially not due to surface roughening, but is due to change in the surface functional groups (as supported by the black coating) which affect the adhesion between rebar and concrete.

The corrosion resistance of steel rebar in concrete greatly affects the durability of steel-reinforced concrete. Water immersion (2 days) and sandblasting are similarly effective for treating steel rebars for the purpose of improving the corrosion resistance of the rebar in the concrete. The increase in corrosion resistance is due to the surface uniformity rendered by either treatment.

Vibration damping is valuable for structures, as it mitigates hazards (whither due to accidental loading, wind, ocean waves or earthquakes), increases the comfort of people who use the structures, and enhances the reliability and performance of structures. Both passive and active methods of damping are useful, although active methods are usually more expensive due to the devices involved. Passive damping most commonly involves the use of viscoelastic materials such as rubber, though these materials tend to suffer from their poor stiffness and high cost compared to the structural material (i.e., concrete). High stiffness is useful for vibration reduction. These problems with stiffness and cost can be removed if the structural material itself has a high damping capacity.

The use of the structural material for passive damping also lowers the cost of damping implementation. Moreover, due to the large volume of structural material in a structure, the resulting damping ability can be substantial. Therefore, the development of concrete that inherently has a high damping capacity is of interest.

The vibration reduction ability of mortar, as expressed by the loss modulus (product of loss tangent and storage modulus) under dynamic flexure (0.2 – 1.0 Hz), is increased by up to 91% by sandblasting the steel rebar, due to the increase in the damping capacity [341]. Surface treatment of the rebar by ozone has negligible effect on the loss modulus.

Sandblasting has a similar effect on the corrosion resistance of steel reinforced concrete as water immersion (by immersion of steel in water to form a surface oxide [340]). It increases the bond strength between steel and concrete than water immersion [340], but less of an increase than from ozone treatment (surface oxidation by exposure to ozone gas) [339]. It increases the damping capacity of steel-reinforced mortar more than ozone treatment [341] and gives a higher storage modulus than ozone treatment [341]. Although ozone treatment has not been evaluated in terms of the corrosion resistance, it is expected to be at least as effective as water immersion, since both treatments involve the formation of a surface oxide film and ozone treatment gives a higher bond strength between steel and concrete than water immersion [339]. Epoxy coating is detrimental to the bond strength, although it enhances the corrosion resistance. All treatments other than sandblasting and ozone treatment have not been evaluated in terms of the damping capacity. With all available data taken into account, it is concluded that sandblasting is, so far, the best steel surface treatment for attaining the combined effects of enhancing the corrosion resistance and damping capacity of steel-reinforced concrete and of increasing the bond strength between steel and concrete.

1.4.3   Admixture surface treatments for interface engineering

Cement-based materials containing solid admixtures such as silica fume and short carbon fibers are improved by surface treatment prior to using the admixtures. Consistency (workability), static and dynamic mechanical properties, specific heat and drying shrinkage are improved [343,344].

Mortar with high consistency, even without a water-reducing agent, is obtained by using silica fume that has been surface-treated with silane, which is hydrophylic [343]. The treatment also increases the strength and modulus, both under tension and compression. In particular, the tensile strength is increased by 31% and the compressive strength is increased by 27%. Moreover, the flexural storage modulus (stiffness), loss tangent (damping capacity) and density are increased [343].

The tensile strength of cement paste is increased by 56%, and the modulus and ductility are increased by 39% by using silane treated carbon fibers and silane-treated silica fume, relative to the values for cement paste with as-received carbon fibers and as-received silica fume [343]. Silane treatment of fibers and silica fume contributes about equally to the strengthening effect. Silane treatment of fibers and silica fume also decreases the air void content. The effects on strengthening and air void content reduction are less when the fiber treatment involves potassium dichromate instead of silane and even less when the treatment involves ozone.

The addition of short carbon fibers to cement paste containing silica fume and methylcellulose causes the loss tangent under flexure (≤ 1 Hz) to decrease by up to 25% and the storage modulus (≤ 2 Hz) to increase by up to 67%, such that both effects increase in the following order: as-received fibers, ozone-treated fibers, dichromate-treated fibers and silane-treated fibers [343]. Silane treatment of silica fume has little effect on the loss tangent, but increases the storage modulus by up to 38% [343].

The specific heat of cement paste is increased by 12% and the thermal conductivity is decreased by 40% by using silane-treated silica fume and silane-treated carbon fibers [343]. The specific heat is increased by the carbon fiber addition, due to fiber-matrix interface slippage. The increase is also in the above order, due to the increasing contribution of the movement of the fiber-matrix covalent coupling. The specific heat is increased by the silica fume addition, due to slippage at the interface between silica fume and cement. The increase is enhanced by silane treatment of the silica fume. Silane treatment of carbon fibers decreases the thermal conductivity.

Silane treatment of carbon fibers and silica fume increase the effectiveness of these admixtures in reducing the drying shrinkage of cement paste [344].

The effects of ozone, dichronate and silane treatments on the surface elemental composition (based on ESCA, i.e., electron spectroscopy for chemical analysis) of carbon fiber are shown in Table 1.10 [344]. The surface carbon concentration is decreased and the surface oxygen concentration is increased by any of the three surface treatments. In the case of the O_3 and dichromate treatments, this is due to the oxidation of the fiber surface and the introduction of hydrophylic functional groups such as –OH and –COOH to the surface. The charge corrected binding energies (C_{1s} and O_{1s}) of both O_3- and dichromate-treated carbon fiber surfaces confirm the existence of these functional groups. In the case of the silane treatment, the oxygen, nitrogen and silicon concentrations on the surface increase and the carbon concentration on the surface decreases after the treatment, due to the composition of the silane coating of the surface (Table 1.10). The charge corrected binding energies of the silane-treated fiber surface confirm the presence of silane.

Table 1.10   Surface elemental composition (in at.%) of carbon fiber

Element	As-received	O_3 treated	Dichromate-treated	Silane-treated
O	13	20.2	20.6	21.8
N	-	-	-	9.3
C	85.4	79.8	79.4	54.3
Cl	-	-	-	4.0
Si	1.7	-	-	10.6

Table 1.11   Surface elemental composition (in at.%) of silica fume

Element	As-received	Silane-treated
O	56.7	50.0
C	10.5	19.3
Si	32.8	30.7
N	-	-

Table 1.11 shows the ESCA results of silica fume particles with and without treatment. Compared to as-received silica fume, the silane-treated silica fume particle surface has more carbon, less oxygen and less silicon atoms. This is consistent with the fact that the surface is partly covered by the silane coating and that the silane used contains C, Si and O atoms.

The ESCA results show that both carbon fiber and silica fume particle surfaces are partly coated by or bonded to silane molecules. Due to the hydrophylic nature of silane, the treated fibers and treated silica fume are expected to be more uniformly distributed in the cement. More importantly, the formation of chemical bonds at which silane serves as bridges between the surface of fiber or silica fume and the cement matrix is expected to make the composite denser and stronger, as shown in the case of silica fume [343]. Therefore, the drying shrinkage strain is decreased by silane treatment of fibers and/or silica fume.

1.4.4 Admixtures for interface engineering

Admixtures (liquids or solids) can be used in cement-matrix composites (whether cement paste, mortar or concrete) for interface engineering.

The use of methylcellulose (a liquid solution) and untreated silica fume (solid particles) as two admixtures in cement paste, relative to the use of untreated silica fume as the sole admixture, increases the loss tangent by up to 50% and decreases the storage modulus by up to 14% [343].

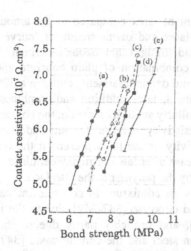

Fig. 1.8   Variation on the contact electrical resistivity with bond strength between steel rebar and concrete at 28 days of curing. (a) Plain concrete and untreated rebar. (b) Concrete with methylcellulose addition and untreated rebar. (c) Concrete with latex addition and untreated rebar. (d) Plain concrete and ozone-treated rebar. (e) Concrete with latex addition and ozone-treated rebar.

The use of silane (a liquid) and untreated silica fume as two admixtures in cement paste, relative to the use of silane-treated silica fume as the sole admixture, increases the compressive modulus, but decreases the compressive ductility and damping capacity [345,346]. It also decreases the air void content and increases the density, specific heat and thermal conductivity. The effects of the silane treatment of silica fume are due to the enhanced hydrophylicity of silica fume and the covalent coupling between silica fume particles and cement. The effects of silane and untreated silica fume as two admixtures are due to the network of covalent coupling among the silica fume particles.

Polymer admixtures (such as methylcellulose and latex) to concrete increase the bond strength between concrete and steel rebar [339]. Fig. 1.8 shows the correlation of the contact resistivity with the bond strength for different polymer admixtures in concrete. Polymer admixtures (curves (b) and (c) of Fig. 1.8) are slightly less effective than ozone treatment of rebar (curve (d) of Fig. 1.8) for increasing the bond strength between rebar and concrete (as well as that between carbon fiber and cement paste). Between the two polymer admixtures, latex (curve (c) of Fig. 1.8) increases the bond strength slightly more significantly than methylcellulose (curve (b) of Fig. 1.8), at least partly due to the large amount of latex compared to the amount of methylcellulose. The combined use of latex and ozone treatment (curve (e) of Fig. 1.8) gives significantly higher bond strength than ozone treatment alone (curve (d) of Fig. 1.8). Relative to the combination of plain concrete and untreated rebar, the combined use of latex and ozone treatment results in a 39% increase in the bond strength. Ozone treatment, latex addition and combined ozone treatment and latex addition cause similarly small increases in the contact resistivity.

The contact resistivity increase after latex addition is presumably due to the high volume resistivity of the latex present at the rebar-concrete interface. The bond strength increase after latex or methylcellulose addition is attributed to the adhesion provided by the polymer at the interface.

Silica fume as an admixture in concrete increases the bond strength between steel rebar and concrete [347-349], due to the increase in the cement matrix modulus [349], and probably partly due to the densification of the transition zone between steel and the cement paste [347,348]. The combined use of silica fume and methylcellulose as two admixtures further enhances the bond strength between rebar and concrete [349] beyond the values attained by

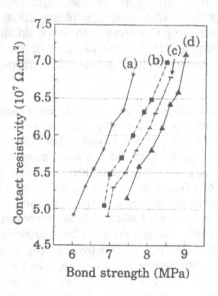

Fig. 1.9   Variation of contact electrical resistivity with bond strength between steel rebar and concrete. (a) Plain concrete. (b) Concrete with silica fume. (c) Concrete with methylcellulose. (d) Concrete with silica fume and methylcellulose.

the use of silica fume as the sole admixture or the use of methylcellulose as the sole admixture [349], as shown in Fig. 1.9.

In spite of the fact that the mechanical interlocking between rebar and concrete due to the surface deformations on the rebar contributes much to the bond strength between rebar and concrete (as shown by the much higher bond strength between rebar and concrete than that between steel fiber and cement paste [350]), the ozone treatment of the rebar and the polymer admixtures to the concrete give significant increases to the bond strength between rebar and concrete. This indicates the importance of interface engineering in improving the bond between rebar and concrete. In the case of the bond between stainless steel fiber and cement paste, the polymer admixtures (latex or methylcellulose) in the cement paste cause the bond strength to increase by 90% [350]. If the surface deformations on the steel rebar were absent, the effects of ozone treatment of rebar and of polymer admixtures in concrete would have been much larger than those described here.

The beneficial effect of polymer admixtures on the bond strength between concrete and concrete has also been shown [351-353]. The presence of a polymer interlayer at the cement-aggregate interface has been shown by microscopy to be responsible for the improved adhesion between cement and aggregate [354].

Admixtures (silica fume and latex) in concrete also enhance the corrosion resistance of steel rebar in the concrete [355]. However, the reason is not related to interface engineering, but is the decrease in the water absorptivity of concrete, and in the case of latex, is also the increase in the electrical resistivity of the concrete.

1.4.5 Conclusion

Interface engineering is effective for improving the mechanical, thermal, chemical (corrosion resistance) and processing (workability) behavior of cement-matrix composites. The techniques of interface engineering include steel rebar surface treatments, admixture surface treatments and the use of admixtures.

1.5   Improving cement-based materials by corrosion protection

1.5.1 Introduction

Steel-reinforced concrete is widely used in construction. The corrosion of the steel-reinforcing bars (rebars) in the concrete limits the life of concrete structures. It is one of the main causes for the deterioration of the civil infrastructure. Corrosion occurs in the steel regardless of the inherent capacity of concrete to protect the steel from corrosion; accelerated corrosion results from the loss of alkalinity in the concrete or the penetration of aggressive ions (such as chloride ions).

Methods of corrosion control of steel reinforced concrete include cathodic protection [352-363], surface treatments of the rebars (epoxy coating [364-395], galvanising [372,383,396-402], copper cladding [403], protective rust growth [404], surface oxidation [405] and sandblasting [405]), the use of admixtures (organic and inorganic corrosion inhibitors [402,406-420], silica fume [422-440], fly ash [441-443], slag [444] and latex [421,445-447]) in the concrete, and the use of surface coating on the concrete [448-450]. This section is a review of the methods and materials for corrosion control of steel-reinforced concrete.

1.5.2   Steel surface treatment

Steel rebars are mostly made of mild steel, due to the importance of low cost. (Stainless steel is excellent in corrosion resistance [451], but its high cost makes it impractical for use in concrete.) The coating of a steel rebar with epoxy (which acts as a barrier) is commonly used to improve the corrosion resistance of the rebar. However, it degrades the bond between rebar and concrete, and the tendency of the epoxy coating to debond is a problem [364-395]. Furthermore, areas of the rebar where the epoxy coating are damaged, and the cut ends of the rebar are not protected from corrosion. On the other hand, galvanized steel attains corrosion protection by its zinc coating, which acts as a sacrificial anode. Galvanized steel tends to bond to concrete better than epoxy coated steel [399], and the tendency of the coating to debond is also lower for galvanized steel. Areas of the rebar where the zinc coating is damaged are still protected; the exposed areas, such as the cut ends, are protected, provided that they are less than 8 mm from the zinc coating [400]. Steel surface treatments that improve both corrosion resistance and bond strength are attractive. They include sandblasting and surface oxidation [405].

Sandblasting involves the blasting of ceramic particles (typically alumina particles of around 250 μm) under pressure (typically around 80 psi or 0.6 MPa). It results in roughening as well as cleaning of the surface of the steel rebar. The cleaning relates to the removal of rust and other contaminants on the rebar surface, as a steel rebar typically is covered by rust and other contaminants here and there. The cleaning causes the surface of the rebar to be more uniform in composition, thus improving the corrosion resistance. The roughening enhances mechanical interlocking between rebar and concrete, thus increasing the bond strength [405].

Water immersion means total immersion of the rebar in water at room temperature for two days. It causes the formation of a black oxide layer on the surface of the rebar, thus enhancing the composition uniformity of the surface and improving the corrosion resistance. In addition, the oxide layer enhances the adhesion between rebar and concrete, thereby increasing the bond strength. Water immersion times that are less than or greater than two days have less desirable effects on both bond strength and corrosion resistance [405].

Steel rebars can also be coated with a corrosion inhibiting cement slurry [402,406,452] or a cement-polymer composite [452] for the purpose of corrosion protection, as described in Section 1.5.3 below.

Of all the methods described above for treating the surface of the steel rebar, the most widely used methods are epoxy coating and galvanizing, due to their relatively long history of usage.

1.5.3   Admixtures in concrete

Admixtures are solids or liquids that are added to a concrete mix to improve the properties of the resulting concrete. Admixtures that enhance the corrosion resistance of steel-reinforced concrete include those that are primarily for corrosion inhibition and those that are primarily for improving the structural properties of concrete. The latter is attractive due to its multifunctionality. The former is mostly inorganic chemicals (such as calcium nitrite [407,418-421], copper oxide [410], zinc oxide [410], sodium thiocyanate [411,412] and alkaline earth silicate [415]) which increase the alkalinity of the concrete, although it can be organic chemicals, such as banana juice [413]. Admixtures that are primarily for structural property improvement can be solid particles, such as silica fume [422-440], fly ash [441-443] and slag [444], and solid particle dispersions, such as latex [421,445-447].

Silica fume (a fine particulate) as an admixture is particularly effective for improving the corrosion resistance of steel-reinforced concrete, due to the decrease in the water absorptivity (or permeability), and not so much due to the increase in electrical resistivity [422-440]. Latex improves the corrosion resistance because it decreases the water absorptivity (or permeability) and increases the electrical resistivity [421,445-447]. Methylcellulose improves the corrosion resistance only slightly [423]. Carbon fibers (short, at a volume fraction below the percolation threshold) decrease the corrosion resistance due to decrease in the electrical resistivity [423]. However, the negative effect of the carbon fibers can be compensated by adding either silica fume or latex, which reduce the water absorptivity [423]. In other words, the corrosion resistance of carbon fiber-reinforced concrete, which typically contains silica fume for improving the fiber dispersion, is superior to that of plain concrete [423].

Table 1.12 [423] shows the effect of silica fume, latex, methylcellulose and short carbon fibers as admixtures on the corrosion potential (E_{corr}, measured according to ASTM C876 using a high impedance voltmeter and a saturated calomel electrode placed on the concrete surface; E_{corr} that is more negative than -270 mV suggests 90% probability of active corrosion) and the corrosion

Table 1.12   Effect of carbon fibers (f), methylcellulose (M), silica fume (SF) and latex (L) on the corrosion resistance of steel rebar in concrete.

	In saturated Ca(OH)$_2$ solution		In 0.5 N NaCl solution	
	E_{corr}^* $\cdot$(-mV, ±5)	I_{corr}^* (μA/cm^2, ±0.03)	E_{corr}^* (-mV, ±5)	I_{corr}^* (μA/cm^2, ±0.03)
P	210	0.74	510	1.50
+ M	220	0.73	/	/
+ M + f	220	0.68	560	2.50
+ M + SF	137	0.17	/	/
+ M + f + SF	170	0.22	350	1.15
+ SF	140	0.19	270	0.88
+ L	180	0.36	360	1.05
+ L + f	190	0.44	405	1.28

Note: P = plain, M = methylcellulose, f = carbon fibers, SF = silica fume, L = latex.
*Value at 25 weeks of corrosion testing.

current density (I_{corr}, determined by measuring the polarization resistance at a low scan rate of 0.167 mV/s) of steel-reinforced concrete in both saturated Ca(OH)$_2$ and 0.5 N NaCl solutions. The saturated Ca(OH)$_2$ solution simulates the ordinary concrete environment; the NaCl solution represents a high chloride environment.   Silica fume improves the corrosion resistance of rebars in concrete in both saturated Ca(OH)$_2$ and NaCl solutions more effectively than any of the other admixtures, although latex is effective.   Methylcellulose improves slightly the corrosion resistance of rebar in concrete in Ca(OH)$_2$ solution. Carbon fibers decrease the corrosion resistance of rebars in concrete, mainly because they decrease the electrical resistivity of concrete. The negative effect of fibers can be compensated by either silica fume or latex.

Instead of using a corrosion inhibiting admixture in the entire volume of concrete, one may use the admixture to modify the cement slurry that is used as a coating on the steel rebar [402,406]. Compared to the use of rebars that have been either epoxy-coated or galvanized, this method suffers from the labor-intensive site-oriented process involved [452]. On the other hand, the use of a shop-coating based on a cement-polymer composite is an emerging alternative [452].

Of all the admixtures described above for improving the corrosion resistance of steel-reinforced concrete, the most widely used ones are calcium nitrite, silica fume and latex.

1.5.4 Surface coating on concrete

Coatings (such as acrylic rubber) can be applied to the concrete surface for the purpose of corrosion control through improving the impermeability [448-450]. However, this method of corrosion control suffers from the poor durability of the coating and the loss of corrosion protection in the areas where the coating is damaged.

1.5.5 Cathodic protection

Cathodic protection is an effective method for corrosion control of steel reinforced concrete [352-362]. This method involves the application of a voltage so as to force electrons to go to the steel rebar, thereby making the steel a cathode. As the voltage needs to be constantly applied, the electrical energy consumption is substantial. This problem can be alleviated by the use of carbon fiber (short) reinforced concrete, as described below.

As the steel rebar is embedded in concrete, the electrons need to go through the concrete in order to reach the rebar. However, concrete is not very electrically conductive. The use of carbon fiber-reinforced concrete for embedding the rebar to be cathodically protected facilitates cathodic protection, as the short carbon fibers enhance the conductivity of the concrete [356].

For directing electrons to the steel-reinforced concrete to be cathodically protected, an electrical contact is needed on the concrete. The electrical contact is electrically connected to the voltage supply. One of the choices of an electrical contact material is zinc, which is a coating deposited on the concrete by thermal spraying. It has a very low volume resistivity (thus requiring no metal mesh embedment), but it suffers from poor wear and corrosion resistance, the tendency to oxidize, high thermal expansion coefficient, and high material and processing costs. Another choice is a conductor-filled polymer [363], which can be applied as a coating without heating, but it suffers from poor wear resistance, a higher thermal expansion coefficient and high material cost. Yet another choice is a metal (e.g., titanium) strip or wire embedded at one end in cement mortar, which is in the form of a coating on the steel-reinforced concrete. The use of carbon fiber-reinforced mortar for this coating facilitates cathodic protection, as it is advantageous to enhance the conductivity of this coating [356].

Due to the decrease in volume electrical resistivity associated with carbon fiber addition (0.35 vol. %) to concrete (embedding steel rebar), concrete containing carbon fibers and silica fume reduces by 18% the driving voltage required for cathodic protection compared to plain concrete, and by 28% compared to concrete with silica fume. Due to the decrease in resistivity associated with carbon fiber addition (1.1 vol. %) to mortar, overlay (embedding titanium wires for electrical contacts to steel-reinforced concrete) in the form of mortar containing carbon fibers and latex reduces by 10% the driving voltage required for cathodic protection, compared to plain mortar overlay. In spite of the low resistivity of mortar overlay with carbon fibers, cathodic protection requires multiple metal electrical contacts embedded in the mortar at a spacing of 11 cm or less [356].

1.5.6   Steel replacement

The replacement of steel rebars by fiber(continuous)-reinforced polymer rebars is an emerging technology which is attractive due to the corrosion resistance of fiber-reinforced polymer [453-458]. However, this technology suffers from the high cost, the poor bonding between concrete and the fiber-reinforced polymer rebar, in addition to the low ductility of the fiber-reinforced polymer.

1.5.7   Conclusion

Methods of corrosion control of steel-reinforced concrete include steel surface treatment, the use of admixtures in concrete, surface coating on concrete and cathodic protection.

References

1.   M.D. Luther and P.A. Smith, *Proc. Eng. Foundation Conf.* 75-106 (1991).
2.   V.M. Malhotra, *Concr. Int.: Design & Construction* 15(4), 23-28 (1993).
3.   M.D. Luther, *Concr. Int.: Design & Construction* 15(4), 29-33 (1993).
4.   J. Wolsiefer, D.R. Morgan, *Concr. Int.: Design & Construction* 15(4), 34-39 (1993).
5.   M.D.A. Thomas, M.H. Shehata, S.G. Shashiprakash, D.S. Hopkins and K. Cail, *Cem. Concr. Res.* 29(8), 1207-1214 (1999).
6.   J. Punkki, J. Golaszewski and O.E. Gjorv, *ACI Mater. J.* 93(5), 427-431 (1996).
7.   R.P. Khatri, V. Sirivivatnanon and W. Gross, *Cem. Concr. Res.* 25(1), 209-220 (1995).
8.   Z. Bayasi and R. Abitaher, *Concr. Int.: Design & Construction* 14(4), 35-37 (1992).
9.   O.E. Gjorv, *Concr. Int.* 20(9), 57-60 (1998).
10.   H. El-Didamony, A. Amer, M. Heikal and M. Shoaib, *Ceramics-Silikaty* 43(1), 29-33 (1999).
11.   H. El-Didamony, A. Amer, M. Heikal, *Ceramics-Silikaty* 42(4), 171-176 (1998).

12.   M. Nehdi, S. Mindess and P.-C. Aitcin, *Cem. Concr. Res.* 28(5), 687-697 (1998).
13.   F.E. Amparano and Y. Xi, *ACI Mater. J.*95(6), 695-703 (1998).
14.   M. Nehdi and S. Mindess, *Transportation Research Record* (1574), 41-48 (1996).
15.   L. Kucharska and M. Moczko, *Adv. Cem. Res.* 6(24), 139-145 (1994).
16.   P. Rougeron and P.-C. *Aitcin, Cem., Concr. & Aggregates* 16(2), 115-124 (1994).
17.   Z. Bayasi and J. Zhou, *ACI Mater. J.* 90(4), 349-356 (1993).
18.   F. Collins and J.G. Sanjayan, *Cem. Concr. Res.* 29(3), 459-462 (1999).
19.   Y. Xu and D.D.L. Chung, *Cem. Concr. Res.* 29(3), 451-453 (1999).
20.   Y. Xu and D.D.L. Chung, *Cem. Concr. Res.* 30(8), 1305-1311 (2000).
21.   Y. Xu and D.D.L. Chung, *Compos. Interfaces* 7(4), 243-256 (2000).
22.   S.Y.N. Chan and X. Ji, *Cem. Concr. Compos.* 21(4), 293-300 (1999).
23.   V.G. Papadakis, *Cem. Concr. Res.* 29(1), 79-86 (1999).
24.   M.D.A. Thomas, K. Cail and R.D. Hooton, *Canadian J. Civil Eng.* 25(3), 391-400 (1998).
25.   H.A. Toutanji, *Adv. Cem. Res.* 10(3), 135-139 (1998).
26.   R. Lewis, *Concrete* 32(5), 19-20,22 (1998).
27.   R. Breitenbuecher, *Mater. Struct.* 31(207), 209-215 (1998).
28.   B. Persson, *Adv. Cem. Based Mater.* 7(3-4), 139-155 (1998).
29.   W.H. Dilger, S.V.K.M. Rao, *Pci J.* 42(4), 82-96 (1997).
30.   K.H. Khayat, M. Vachon and M.C. Lanctot, *ACI Mater. J.* 94(3), 183-192 (1997).
31.   S.A. El-Desoky I.A. Ibrahim, *European Journal of Control* 43(5), 919-932 (1996).
32.   M.L. Allan and L.E. Kukacka, *ACI Mater. J.* 93(6), 559-568 (1996).
33.   Y. Li, B.W. Langan and M.A. Ward, *Cem., Concr. & Aggregates* 18(2), 112-117 (1996).
34.   M.N. Haque, *Cem. Concr. Compos.* 18(5), 333-342 (1996).
35.   K. Wiegrink, S. Marikunte and S.P. Shah, *ACI Mater. J.* 93(5), 409-415 (1996).
36.   M. Kessal, P.-C. Nkinamubanzi, A. Tagnit-Hamou, P.-C. Aitcin, *Cem., Concr. & Aggregates* 18(1), 49-54 (1996).
37.   V. Lilkov and V. Stoitchkov, *Cem. Concr. Res.* 26(7), 1073-1081 (1996).
38.   M. Kessal, M. Edwards-Lajnef, A. Tagnit-Hamou and P.-C. Aitcin, *Canadian J. Civil Eng.* 23(3), 614-620 (1996).
39.   W.H. Dilger, A. Ghali and S.V.K.M. Rao, *Pci J.* 41(2), 68-89 (1996).
40.   S.A.A. El-Enein, M.F. Kotkata, G.B. Hanna, M. Saad M.M.A. El Razek, *Cem. Concr. Res.* 25(8), 1615-1620 (1995).
41.   S. Wild, B.B. Sabir and J.M. Khatib, *Cem. Concr. Res.* 25(7), 1567-1580 (1995).

42.   K.G. Babu and P.V.S. Prakash, *Cem. Concr. Res.* 25(6), 1273-1283 (1995).
43.   M. Collepardi, S. Monosi and P. Piccioli, *Cem. Concr. Res.* 25(5), 961-968 (1995).
44.   J.C. Walraven, *Betonwerk und Fertigteil-Technik* 60(11), 7 pp (1994).
45.   F. Papworth and R. Ratcliffe, *Concr. Int.* 16(10), 39-44 (1994).
46.   K. Torii and M. Kawamura, *Cem. Concr. Compos.* 16(4), 279-286 (1994).
47.   C. Ozyildirim and W.J. Halstead, *ACI Mater. J.* 91(6), 587-594 (1994).
48.   S.A. Khedr and M.N. Abou-Zeid, *J. Mater. Civil Eng.* 6(3), 357-375 (1994).
49.   P. Fidiestol, *Concr. Int.: Design & Construction* 15(11), 33-36 (1993).
50.   L. Rocole, *Aberdeen's Concrete Construction* 38(6), 441-442 (1993).
51.   G. Ozyildirim, *Concr. Int.: Design & Construction* 15(1), 33-38 (1993).
52.   M. Moukwa, B.G. Lewis, S.P. Shah and C. Ouyang, *Cem. Concr. Res.* 23(3), 711-723 (1993).
53.   B. Ma, J. Li and J. Peng, *J. Wuhan University of Technology*, Materials Science Edition 14(2), 1-7 (1999).
54.   K. Tan and X. Pu, *Cem. Concr. Res.* 28(12), 1819-1825 (1998).
55.   L. Bagel, *Cem. Concr. Res.* 28(7), 1011-1020 (1998).
56.   C.E.S. Tango, *Cem. Concr. Res.* 28(7), 969-983 (1998).
57.   M. Lachemi, G. Li, A. Tagnit-Hamou and P.-C. Aitcin, *Concr. Int.* 20(1), 59-65 (1998).
58.   S.H. Alsayed, *ACI Mater. J.* 94(6), 472-477 (1997).
59.   A. Salas, R. Gutierrez and S. Delvasto, *J. Resource Management & Tech.* 24(2), 74-78 (1997).
60.   J. Li and T. Pei, *Cem. Concr. Res.* 27(6), 833-837 (1997).
61.   B.B. Sabir, *Magazine of Concr. Res.* 49(179), 139-146 (1997).
62.   V. Waller, P. Naproux and F. Larrard, *Bulletin de Liaison des Laboratoires des Ponts et Chaussees* (208), 53-65 (1997).
63.   H.A. Toutanji and T. El-Korchi, *Cem., Concr. & Aggregates* 18(2), 78-84 (1996).
64.   S. Iravani, *ACI Mater. J.* 93(5), 416-426 (1996).
65.   M.H. Zhang, R. Lastra and V.M. Malhotra, Cem. Concr. Res. 26(6), 963-977 (1996).
66.   R. Gagne, A. Boisvert and M. Pigeon, *ACI Mater. J.* 93(2), 111-120 (1996).
67.   S.L. Mak and K. Torii, *Cem. Concr. Res.* 25(8), 1791-1802 (1995).
68.   B.B. Sabir, *Magazine of Concr. Res.* 47(172), 219-226 (1995).
69.   H.A. Toutanji and T. El-Korchi, *Cem. Concr. Res.* 25(7), 1591-1602 (1995).

70. S. Ghosh and K.W. Nasser, *Canadian J. Civil Eng.* 22(3), 621-636 (1995).
71. H. Marzouk and A. Hussein, *J. Mater. Civil Eng.* 7(3), 161-167 (1995).
72. E.H. Fahmy, Y.B.I. Shaheen and W.M. El-Dessouki, *J. Ferrocement* 25(2), 115-121 (1995).
73. F.P. Zhou, B.I.G. Barr and F.D. Lydon FD, *Cem. Concr. Res.* 25(3), 543-552 (1995).
74. J. Xie, A.E. Elwi and J.G. MacGregor, *ACI Mater. J.* 92(2), 135-145 (1995).
75. S.U. Al-Dulaijan, A.H.J. Al-Tayyib, M.M. Al-Zahrani, G. Parry-Jones and A.I. Al-Mana, *J. American Ceramic Soc.* 78(2), 342-346 (1995).
76. A. Goldman and A. Bentur, *Adv. Cem. Based Mater.* 1(5), 209-215 (1994).
77. A. Goldman and A. Bentur, *Cem. Concr. Res.* 23(4), 962-972 (1993).
78. K.J. Folliard, M. Ohta, E. Rathje and P. Collins, *Cem. Concr. Res.* 24(3), 424-432 (1994).
79. C. Xiaofeng, G. Shanglong, D. Darwin and S.L. McCabe, *ACI Mater. J.* 89(4), 375-387 (1992).
80. R.S. Ravindrarajah, *Nondestructive Testing of Concrete Elements and Structures*, Proc. Nondestr. Test. Concr. Elem. Struct., (1992), ASCE, NewYork, NY, pp. 115-126.
81. B.B. Sabir and K. Kouyiali, *Cem. Concr. Res.* 13(3), 203-208 (1991).
82. H.A. Toutanji, L. Liu and T. El-Korchi, *Mater. Struct.* 32(217), 203-209 (1999).
83. S. Sarkar, O. Adwan and J.G.L. Munday, *Struct. Eng.* 75(7), 115-121 (1997).
84. C. Tasdemir, M.A. Tasdemir, N. Mills, B.I.G. Barr and F.D. Lydon, *ACI Mater. J.* 96(1), 74-83 (1999).
85. M.G. Alexander and T.I. Milne, *ACI Mater. J.* 92(3), 227-235 (1995).
86. X. Fu, X. Li and D.D.L. Chung, *J. Mater. Sci.* 33, 3601-3605 (1998).
87. Y. Wang and D.D.L. Chung, *Cem. Concr. Res.* 28(10), 1353-1356 (1998).
88. M. Tamai and M. Tanaka, *Transactions of the Japan Concrete Institute* 16, 81-88 (1994).
89. R. Quaresima, G. Scoccia, R. Volpe, F. Medici and C. Merli, *Proc. 1993 Symp. Stabilization and Solidification of Hazardous, Radioactive, and Mixed Wastes*: 3rd Volume, ASTM Special Technical Publication, vol. 1240, (1996), ASTM, Conshohocken, PA, pp. 135-146.
90. M. Sonebi and K.H. Khayat, *Canadian J. Civil Eng.* 20(4), 650-659 (1993).
91. R.D. Hooton, *ACI Mater. J.* 90(2), 143-151 (1993).
92. M. Lachemi, G. Li, A. Tagnit-Hamou and P.-C. Aitcin, *Concr. Int.* 20(1), 59-65 (1998).

93.   V. Yogendran, B.W. Langan, M.N. Haque and M.A. Ward, *ACI Mater. J.* 84(2), 124-129 (1987).
94.   P.-W. Chen and D.D.L. Chung, *Composites* 24(1), 33-52 (1993).
95.   M.G. Kashi and R.E. Weyers, Proc. Sessions related to Structural Materials at Structures Congress, vol. 89, (1989) ASCE, New York, NY, pp. 138-148.
96.   V.M. Malhotra, *Fly Ash, Silica Fume, Slag, and Natural Pozzolans in Concrete*, Proc. 2nd Int. Conf., Publication SP – American Concrete Institute 91, vol. 2, (1986), American Concr. Inst., Detroit, MI, pp. 1069-1094.
97.   A. Durekovic, V. Calogovic and K. Popovic, *Cem. Concr. Res.* 19(2), 267-277 (1989).
98.   B.B. Sabir, *Cem. Concr. Compos.* 19(4), 285-294 (1997).
99.   Z.-Q. Shi and D.D.L. Chung, *Cem. Concr. Res.* 27(8), 1149-1153 (1997).
100.   X. Li and D.D.L. Chung, *Cem. Concr. Res.* 28(4), 493-498 (1998).
101.   V. Baroghel-Bouny and J. Godin, *Bulletin de Liaison des Laboratoires des Ponts et Chaussees* (218), 39-48 (1998).
102.   S.H. Alsayed, *Cem. Concr. Res.* 28(10), 1405-1415 (1998).
103.   A. Lamontagne, M. Pigeon, R. Pleau and D. Beaupre, *ACI Mater. J.* 93(1), 69-74 (1996).
104.   M.G. Alexander, *Adv. Cem. Res.* 6(22), 73-81 (1994).
105.   F.H. Al-Sugair, *Magazine Concr. Res.* 47(170), 77-81 (1995).
106.   B. Bissonnette and M. Pigeon, *Cem. Concr. Res.* 25(5), 1075-1085 (1995).
107.   G.A. Rao, *Cem. Concr. Res.* 28(10), 1505-1509 (1995).
108.   Z. Li, M. Qi, Z. Li and B. Ma, *J. Mater. Civil Eng.* 11(3), 214-223 (1999).
109.   O.M. Jensen and P.F. Hansen, *ACI Mater. J.* 93(6), 539-543 (1996).
110.   O.M. Jensen and P.F. Hansen, *Adv. Cem. Res.* 7(25), 33-38 (1995).
111.   B. Persson, *Cem. Concr. Res.* 28(7), 1023-1036 (1998).
112.   T.A. Samman, W.H. Mirza and F.F. Wafa, *ACI Mater. J.* 93(1), 36-40 (1996).
113.   R. Bloom and A. Bentur, *ACI Mater. J.* 92(2), 211-217 (1995).
114.   X. Fu and D.D.L. Chung, *ACI Mater. J.* 96(4), 455-461 (1999).
115.   Z. Liu and J.J. Beaudoin, *Cem. Concr. Res.* 29(7), 1085-1090 (1999).
116.   T.-J. Zhao, J.-Q. Zhu and P.-Y. Chi, *ACI Mater. J.* 96(1), 84-89 (1999).
117.   T.H. Wee, A.K. Suryavanshi and S.S. Tin, *Cem. Concr. Compos.* 21(1), 59-72 (1999).
118.   R.K. Dhir and M.R. Jones, *Fuel* 78(2), 137-142 (1999).
119.   P. Sandberg, L. Tang and A. Anderson, *Cem. Concr. Res.* 28(10), 1489-1503 (1998).

120. N. Gowripalan and H.M. Mohamed, *Cem. Concr. Res.* 28(8), 1119-1131 (1998).
121. C. Shi, J.A. Stegemann and R.J. Caldwell, *ACI Mater. J.* 95(4), 389-394 (1998).
122. S.L. Amey, D.A. Johnson, M.A. Miltenberger and H. Farzam, *ACI Struct. J.* 95(2), 205-214 (1998).
123. R.J. Detwiler, T. Kojundic and P. Fidjestol, Concr. Int. 19(8), 43-45 (1997).
124. S.L. Amey, D.A. Johnson, M.A. Miltenberger and H. Farzam, *ACI Struct. J.* 95(1), 27-36 (1997).
125. R.P. Khatri, V. Sirivivatnanon and L.K. Yu, *Magazine Concr. Res.* 49(180), 167-172 (1997).
126. R.P. Khatri, V. Sirivivatnanon and J.L. Yang, *Cem. Concr. Res.* 27(8), 1179-1189 (1997).
127. K. Tan and O.E. Gjorv, *Cem. Concr. Res.* 26(3), 355-361 (1996).
128. G.J.Z. Xu, D.F. Watt and P.P. Hudec, *Cem. Concr. Res.* 25(6), 1225-1236 (1995).
129. A.A. Ramezanianpour and V.M. Malhotra, *Cem. Concr. Compos.* 17(2), 125-133 (1995).
130. O.S.B. Al-Amoudi, M. Maslehuddin and Y.A.B. Abdul-Al, *Construction & Building Mater.* 9(1), 25-33 (1995).
131. A. Delagrave, M. Pigeon and E. Revertegat, *Cem. Concr. Res.* 24(8), 1433-1443 (1994).
132. C. Ozyildirim, *Cem., Concr. & Aggregates* 16(1), 53-56 (1994).
133. C. Ozyildirim, *ACI Mater. J.* 91(2), 197-202 (1994).
134. R.J. Detwiler, C.A. Fapohunda and J. Natale, *ACI Mater. J.* 91(1), 63-66 (1994).
135. C. Lobo and M.D. Cohen, *ACI Mater. J.* 89(5), 481-491 (1992).
136. N. Banthia, M. Pigeon, J. Marchand and J. Boisvert, *J. Mater. Civil Eng.* 4(1), 27-40 (1992).
137. B. Ma, Z. Li and J. Peng, *J. Wuhan Univ. of Technology*, Materials Science Edition 13(4), 16-24 (1998).
138. P.S. Mangat and B.T. Molloy, *Magazine Concr. Res.* 47(171), 129-141 (1995).
139. C.S. Poon, L. Lam and Y.L. Wong, *J. Mater. Civil Eng.* 11(3), 197-205 (1999).
140. H. Yan, W. Sun and H. Chen, *Cem. Concr. Res.* 29(3), 423-426 (1999).
141. D.R.G. Mitchell, I. Hinczak and R.A. Day, *Cem. Concr. Res.* 28(11), 1571-1584 (1998).
142. K.O. Kjellsen, O.H. Wallevik and L. Fjallberg, *Adv. Cem. Res.* 10(1), 33-40 (1998).
143. K. Vivekanandam and I. Patnaikuni, *Cem. Concr. Res.* 27(6), 817-823 (1997).

144.   M. Saad, S.A. Abo-El-Enein, G.B. Hanna and M.F. Kotkata, *Cem. Concr. Res.* 26(10), 1479-1484 (1996).

145.   K.A. Khalil, *Mater. Letters* 26(4-5), 259-264 (1996).

146.   J. Marchand, H. Hornain, S. Diamond, M. Pigeon and H. Guiraud, *Cem. Concr. Res.* 26(3), 427-438 (1996).

147.   C. Tasdemir, M.A. Tasdemir, F.D. Lydon and B.I.G. Barr, *Cem. Concr. Res.* 26(1), 63-68 (1996).

148.   M.J. Aquino, Z. Li and S.P. Shah, *Adv. Cem. Based Mater.* 2(6), 211-223 (1995).

149.   M. Cheyrezy, V. Maret and L. Frouin, *Cem. Concr. Res.* 25(7), 1491-1500 (1995).

150.   D.P. Bentz and P.E. Stutzman, *Cem. Concr. Res.* 24(6), 1044-1050 (1994).

151.   Y. Cao and R.J. Detwiler, *Cem. Concr. Res.* 25(3), 627-638 (1995).

152.   P.-C. Aitcin, *Construction & Building Mater.* 9(1), 13-17 (1995).

153.   A. Durekovic, *Cem. Concr. Res.* 25(2), 365-375 (1995).

154.   S. Mindess, L. Qu and M.G. Alexander, *Adv. Cem. Res.* 6(23), 103-107 (1994).

155.   Q. Yu, M. Hu, J. Qian, X. Wang and C. Tao, *J. Wuhan Univ. of Technology* 9(2), 22-28 (1994).

156.   K. Mitsui, Z. Li, D.A. Lange and S.P. Shah, *ACI Mater. J.* 91(1), 30-39 (1994).

157.   M.D. Cohen, A. Goldman and W.-F. Chen, *Cem. Concr. Res.* 24(1), 95-98 (1994).

158.   D.N. Winslow, M.D. Cohen, D.P. Bentz, K.A. Snyder and E.J. Garboczi, *Cem. Concr. Res.* 24(1), 25-37 (1994).

159.   Z. Xu, P. Gu, P. Xie J.J. Beaudoin,, *Cem. Concr. Res.* 23(5), 1007-1015 (1993).

160.   S.L. Sarkar, S. Chandra and L. Berntsson, *Cem. Concr. Compos.* 14(4), 239-248 (1992).

161.   D.P. Bentz, P.E. Stutzman and E.J. Garboczi, *Cem. Concr. Res.* 22(5), 891-902 (1992).

162.   X. Ping and J.J. Beaudoin, *Cem. Concr. Res.* 22(4), 597-604 (1992).

163.   S.P. Shah, Z. Li and D.A. Lange, *Proc. 9th Conf. Eng. Mechanics*, (1992), New York, NY, pp. 852-855.

164.   J. Zemajtis, R.E. Weyers and M.M. Sprinkel, *Transportation Res. Record 1999* pp. 57-59.

165.   O.E. Gjorv, *ACI Mater. J.* 92(6), 591-598 (1995).

166.   J.G. Cabrera, P.A. Claisse and D.N. Hunt, *Construction & Building Mater,* 9(2), 105-113 (1995).

167.   N.R. Jarrah, O.S.B. Al-Amoudi, M. Maslehuddin, O.A. Ashiru and A.I. Al-Mana, *Construction & Building Mater.* 9(2), 97-103 (1995).

168.   T. Lorentz and C. French, ACI Mater. J. 92(2), 181-190 (1995).

169. S.A. Khedr and A.F. Idriss, *J. Mater. Civil Eng.* 7(2), 102-107 (1995).
170. P.S. Mangat, J.M. Khatib and B.T. Molloy, *Cem. Concr. Compos.* 16(2), 73-81 (1994).
171. O.S.B. Al-Amoudi, Rasheeduzzafar, M. Maslehuddin and S.N. Abduljauwad, *Cem., Concr. & Aggregates* 16(1), 3-11 (1994).
172. D. Whiting, *Transportation Res. Record* (1392), 142-148 (1993).
173. J.T. Wolsiefer Sr., *Utilization of Industrial By-Products for Construction Materials*, Proc. ASCE Natl. Conv. Expo., (1993), ASCE, New York, NY, pp. 15-29.
174. Rasheeduzzafar, S.S. Al-Saadoun and A.S. Al-Gahtani, *ACI Mater. J.* 89(4), 337-366 (1992).
175. Rasheeduzzafar, *ACI Mater. J.* 89(6), 574-586 (1992).
176. C.K. Nmai, S.A. Farrington and G.S. Bobrowski, *Concr. Int.: Design & Construction* 14(4), 45-51 (1992).
177. J. Hou and D.D.L. Chung, *Corrosion Sci.* 42(9), 1489-1507 (2000).
178. V.S. Ramachandran, *Cem. Concr. Compos.* 20(2/3), 149-161 (1998).
179. S. Diamond, *Cem. Concr. Compos.* 19(5-6), 391-401 (1997).
180. H. Wang and J.E. Gillott, *Magazine Concr. Res.* 47(170), 69-75 (1995).
181. W. Wieker, R. Herr and C. Huebert, *Betonwerk und Fertigteil-Technik* 60(11), 86-91 (1994).
182. M. Berra, T. Mangialardi and A.E. Paolini, *Cem. Concr. Compos.* 16(3), 207-218 (1994).
183. M. Geiker and N. Thaulow, *Proc. Eng. Foundation Conf. 1991* pp. 123-136 (1991)
184. A. Shayan, G.W. Quick and C.J. Lancucki, *Adv. Cem. Res.* 5(20), 151-162 (1993).
185. J. Duchesne and M.A. Berube, *Cem. Concr. Res.* 24(2), 221-230 (1994).
186. P.P. Hudec and N.K. Banahene, *Cem. Concr. Compos.* 15(1-2), 21-26 (1993).
187. J.E. Gillot and H. Wang, *Cem. Concr. Res.* 23(4), 973-980 (1993).
188. N.R. Swamy and R.M. Wan, *Cem., Concr. & Aggregates* 15(1), 32-49 (1993).
189. D.S. Lane, *Materials: Performance and Prevention of Deficiencies and Failures*, Mater. Eng. Congr., (1992), ASCE, New York, NY, pp. 231-244.
190. Rasheeduzzafar and S.E. Hussain, *Cem. Concr. Compos.* 13(3), 219-225 (1991).
191. C. Perry and J.E. Gillott, *Durability of Building Mater.* 3(2), 1985 (1985).
192. Rasheeduzzafar and H.S. Ehtesham, *Cem. & Concr. Compos.* 13(3), 219-225 (1991).

193.   B. Durand, J. Berard and R. Roux, *Cem. Concr. Res.* 20(3), 419-428 (1990).

194.   G.J.Z. Xu, D.F. Watt and P.P. Hudec, *Cem. Concr. Res.* 25(6), 1225-1236 (1995).

195.   M.-A. Berube, J. Duchesne and D. Chouinard, *Cem., Concr. & Aggregates* 17(1), 26-33 (1995).

196.   S. Diamond, *Cem. & Concr. Compos.* 19(5-6), 391-401 (1997).

197.   A.H. Ali, *Corrosion Prevention & Control* 46(3), 76-81 (1999).

198.   M.S. Morsy, *Cem. Concr. Res.* 29(4), 603-606 (1999).

199.   R.J. van Eijk and H.J.H. Brouwers, *Heron* 42(4), 215-229 (1997).

200.   A.K. Tamimi, *Mater. Struct.* 30(197), 188-191 (1997).

201.   F. Turker, F. Akoz, S. Koral and N. Yuzer, *Cem. Concr. Res.* 27(2), 205-214 (1997).

202.   F.M. Kilinckale, *Cem. Concr. Res.* 27(12), 1911-1918 (1997).

203.   N.D. Belie, V.D. Coster and D.V. Nieuwenburg, *Magazine Concr. Res.* 49(181), 337-344 (1997).

204.   P. Jones, *Concrete* (London) 31(4), 12-13 (1997).

205.   J.A. Daczko, D.A. Johnson and S.L. Amey, *Mater. Performance* 36(1), 51-56 (1997).

206.   A. Delagrave, M. Pigeon, J. Marchand and E. Revertagat, *Cem. Concr. Res.* 26(5), 749-760 (1996).

207.   A.A. Bubshait, B.M. Tahir and M.O. Jannadi, *Building Res. & Info.* 24(1), 41-49 (1996).

208.   F. Akoz, F. Turker, S. Koral and N. Yuzer, *Cem. Concr. Res.* 25(6), 1360-1368 (1995).

209.   K. Torii and M. Kawamura, *Cem. Concr. Res.* 24(2), 361-370 (1994).

210.   H.-S. Shin and K.-S. Jun, *J. Environmental Sci. & Health*, Part A: Environmental Science & Engineering & Toxic & Hazardous Substance Control 30(3), 651-668 (1995).

211.   V. Matte and M. Moranville, *Cem. Concr. Compos.* 21(1), 1-9 (1999).

212.   M.J. Rudin, Waste Management 16(4), 305-311 (1996).

213.   P.-W. Chen and D.D.L. Chung, *Compos.*: Part B 27B, 269-274 (1996).

214.   A. Yahia, K.H. Khayat and B. Benmokrane, *Mater. Struct.* 31(208), 267-274 (1998).

215.   B.S. Hamad and S.M. Sabbah, *Mater. Struct.* 31(214), 707-713 (1998).

216.   Z. Li, M. Xu and N.C. Chui, *Magazine Concr. Res.* 50(1), 49-57 (1998).

217.   A. Mor, *ACI Mater. J.* 89(1), 76-82 (1992).

218.   X. Fu and D.D.L. Chung, *ACI Mater. J.* 95(6), 725-734 (1998).

219.   P.J.M. Monteiro, O.E. Gjorv and P.K. Mehta, *Cem. Concr. Res.* 19(1), 114-123 (1989).

220.   O.E. Gjorv, P.J.M. Monteiro and P.K. Mehta, *ACI Mater. J.* 87(6), 573-580 (1990).

221.   X. Fu and D.D.L. Chung, *ACI Mater. J.* 95(5), 601-608 (1998).
222.   K. Kovler, S. Igarashi and A. Bentur, *Mater. Struct.* 32(219), 383-387 (1999).
223.   Y. Xu and D.D.L. Chung, *Cem. Concr. Res.* 29(7), 1117-1121 (1999).
224.   S.B. Park, E.S. Yoon and B.I. Lee, *Cem. Concr. Res.* 29(2), 193-200 (1999).
225.   M. Pigeon and R. Cantin, *Cem. Concr. Compos.* 20(5), 365-375 (1998).
226.   H. Toutanji, S. McNeil and Z. Bayasi, *Cem. Concr. Res.* 28(7), 961-968 (1998).
227.   T.-J. Kim and C.-K. Park, *Cem. Concr. Res.* 28(7), 955-960 (1998).
228.   H.A.D. Kirsten, *J. South African Inst. of Mining & Metallurgy* 98(2), 93-104 (1998).
229.   C. Aldea, S. Marikunte and S.P. Shah, *Adv. Cem. Based Mater.* 8(2), 47-55 (1998).
230.   A. Dubey and N. Banthia, *ACI Mater. J.* 95(3), 284-292 (1998).
231.   L. Biolzi, G.L. Guerrini and G. Rosati, *Construction & Building Mater.* 11(1), 57-63 (1997).
232.   O. Eren and T. Celik, *Construction & Building Mater.* 11(7-8), 373-382 (1997).
233.   S.A. Austin, C.H. Peaston and P.J. Robins, *Construction & Building Mater.* 11(5-6), 291-298 (1997).
234.   E. Dallaire, P.-C. Aitcin and M. Lachemi, *Civil Eng.* (New York) 68(1), 48-51 (1998).
235.   R.J. Brousseau and G.B. Pye, *ACI Mater. J.* 94(4), 306-310 (1997).
236.   S. Marikunte, C. Aldea and S.P. Shah, *Adv. Cem. Based Mater.* 5(3-4), 100-108 (1997).
237.   Y.-W. Chan and V.C. Li, *Adv. Cem. Based Mater.* 5(1), 8-17 (1997).
238.   R. Cantin and M. Pigeon, *Cem. Concr. Res.* 26(11), 1639-1648 (1996).
239.   M.R. Taylor, F.D. Lydon and B.I.G. Barr, *Construction & Building Mater.* 10(6), 445-450 (1996).
240.   H.-C. Wu, V.C. Li, Y.M. Lim, K.F. Hayes and C.C. Chen, *J. Mater. Sci. Lett.* 15(19), 1736-1739 (1996).
241.   A. Kumar and A.P. Gupta, *Experimental Mechanics* 36(3), 258-261 (1996).
242.   S. Wei, G. Jianming and Y. Yun, *ACI Mater. J.* 93(3), 206-212 (1996).
243.   N. Banthia and C. Yan, *Cem. Concr. Res.* 26(5), 657-662 (1996).
244.   A. Katz and A. Bentur, *Adv. Cem. Based Mater.* 3(1), 1-13 (1996).
245.   P. Balaguru and A. Foden, *ACI Struct. J.* 93(1), 62-78 (1996).
246.   P. Soroushian, F. Mirza and A. Alhozaimy, *ACI Mater. J.* 92(3), 291-295 (1995).
247.   A. Katz and A. Bentur, *Cem. Concr. Compos.* 17(2), 87-97 (1995).
248.   N. Banthia, N. Yan, C. Chan, C. Yan and A. Bentur, *Microstructrure of Cement-Based Systems/Bonding and Interfaces in Cementitious*

Materials, Materials Research Society Symposium Proceedings, vol. 370, (1995), Materials Research Society, Pittsburgh, PA, pp. 539-548.

249.   A. Katz and V.C. Li, *Microstructure of Cement-Based Systems/Bonding and Interfaces in Cementitious Materials*, Materials Research Society Symposium Proceedings, vol. 370, (1995), Materials Research Society, Pittsburgh, PA, pp. 529-537.

250.   A. Katz, V.C. Li, and A. Kazmer, *J. Mater. Civil Eng.* 7(2), 125-128 (1995).

251.   W.M. Boone, D.B. Clark and E.L. Theisz, *Proc. Ports '95 Conference on Port Engineering and Development for the 21st Century*, vol. 2, (1995), ASCE, New York, NY, pp. 1138-1147.

252.   S.-I. Igarashi and M. Kawamura, *Doboku Gakkai Rombun-Hokokushu/Proceedings of the Japan Society of Civil Engineers* (502),pt. 5-25, 83-92 (1994).

253.   S. Marikunte and P. Soroushian, *ACI Mater. J.* 91(6), 607-616 (1994).

254.   X. Lin, M.R. Silsbee, D.M. Roy, K. Kessler and P.R. Blankenhorn, Cem. Concr. Res. 24(8), 1558-1566 (1994).

255.   Anonymous, *Enr* (Engineering News-Record) 233(2), 13 (1994).

256.   N.M.P. Low and J.J. Beaudoin, *Cem. Concr. Res.* 24(5), 874-884 (1994).

257.   N.M.P. Low and J.J. Beaudoin, *Cem. Concr. Res.* 24(2), 250-258 (1994).

258.   A. Katz and A. Bentur, *Cem. Concr. Res.* 24(2), 214-220 (1994).

259.   P. Balaguru and M.G. Dipsia, *ACI Mater. J.* 90(5), 399-405 (1993).

260.   N.M.P. Low and J.J. Beaudoin, *Cem. Concr. Res.* 23(6), 1467-1479 (1993).

261.   P.S. Mangat and G.S. Manarakis, *Mater. Struct.* 26(161), 433-440 (1993).

262.   N.M.P. Low and J.J. Beaudoin, *Cem. Concr. Res.* 23(5), 1016-1028 (1993).

263.   N.M.P. Low and J.J. Beaudoin, *Cem. Concr. Res.* 23(4), 905-916 (1993).

264.   S.B. Park and B.I. Lee, *High Temperatures – High Pressures* 22(6), 663-670 (1990).

265.   P. Balaguru, R. Narahari and M. Patel, *ACI Mater. J.* 89(6), 541-546 (1992).

266.   A.S. Ezeldin and P.N. Balaguru, *J. Mater. Civil Eng.* 4(4), 415-429 (1992).

267.   N.M.P. Low and J.J. Beaudoin, *Cem. Concr. Res.* 22(5), 981-989 (1992).

268.   P.-W. Chen, X. Fu and D.D.L. Chung, *ACI Mater. J.* 94(2), 147-155 (1997).

269.   H. Yan, W. Sun and H. Chen, *Cem. Concr. Res.* 29(3), 423-426 (1999).

270. J. Cao and D.D.L. Chung, *Cem. Concr. Res.* 31(11), 1633-1637 (2001).

271. J. Cao and D.D.L. Chung, *Cem. Concr. Res.* 31(11), 379-385 (2002).

272. P. Gu and J.J. Beaudoin, *Adv. Cem. Res.* 9(33), 1-8 (1997).

273. S. Wen and D.D.L. Chung, *Cem. Concr. Res.* 31(4), 673-677 (2001).

274. J.W. Newman, *Int. SAMPE Symp. Exhib.*, vol. 32, (1987), SAMPE, Covina, CA, pp. 938-944.

275. S. Furukawa, Y. Tsuji and S. Otani, *Proc. 30th Japan Congress on Materials Research*, (1987), Soc. of Materials Science, Kyoto, Jpn., pp. 149-152.

276. K. Saito, N. Kawamura and Y. Kogo, *Advanced Materials: The Big Payoff*, National SAMPE Technical Conf., vol. 21, (1989), Covina, CA, pp. 796-802.

277. S. Wen and D.D.L. Chung, *Cen. Concr. Res.* 29(3), 445-449.

278. T. Sugama, L.E. Kukacka, N. Carciello and D. Stathopoulos, *Cem. Concr. Res.* 19(3), 355-365 (1989).

279. X. Fu, W. Lu and D.D.L. Chung, *Cem. Concr. Res.* 26(7), 1007-1012 (1996).

280. X. Fu, W. Lu and D.D.L. Chung, *Carbon* 36(9), 1337-1345 (1998).

281. Y. Xu and D.D.L. Chung, *Cem. Concr. Res.* 29(5), 773-776 (1999).

282. T. Yamada, K. Yamada, R. Hayashi and T. Herai, *Int. SAMPE Symp. Exhib.*, vol. 36, (1991), SAMPE, Covina, CA, pt. 1, pp. 362-371.

283. T. Sugama, L.E. Kukacka, N. Carciello and B. Galen, *Cem. Concr. Res.* 18(2), 290-300 (1988).

284. B.K. Larson, L.T. Drzal and P. Sorousian, *Composites* 21(3), 205-215 (1990).

285. A. Katz, V.C. Li and A. Kazmer, *J. Materials Civil Eng.* 7(2), 125-128 (1995).

286. S.B. Park and B.I. Lee, *Cem. Concr. Compos.* 15(3), 153-163 (1993).

287. P. Chen, X. Fu and D.D.L. Chung, *ACI Mater. J.* 94(2), 147-155 (1997).

288. P. Chen and D.D.L. Chung, *Composites* 24(1), 33-52 (1993).

289. A.M. Brandt and L. Kucharska, *Materials for the New Millennium,* Proc. Mater. Eng. Conf., vol. 1, (1996), ASCE, New York, NY, pp. 271-280.

290. H.A. Toutanji, T. El-Korchi, R.N. Katz and G.L. Leatherman, *Cem. Concr. Res.* 23(3), 618-626 (1993).

291. N. Banthia and J. Sheng, *Cem. Concr. Composites* 18(4), 251-269 (1996).

292. H.A. Toutanji, T. El-Korchi and R.N. Katz, *Com. Con. Compos.* 16(1), 15-21 (1994).

293. S. Akihama, T. Suenaga and T. Banno, *Int. J. Cem. Compos. & Lightweight Concr.* 6(3), 159-168 (1984).

294.   M. Kamakura, K. Shirakawa, K. Nakagawa, K. Ohta and S. Kashihara, *Sumitomo Metals* (1983).

295.   A. Katz and A. Bentur, *Cem. Concr. Res.* 24(2), 214-220 (1994).

296.   Y. Ohama and M. Amano, *Proc. 27ᵗʰ Japan Congress on Materials Research,* (1983), Soc. Mater. Sci., Kyoto, Japan, pp. 187-191.

297.   Y. Ohama, M. Amano and M. Endo, *Concrete Int.: Design & Construction,* 7(3), 58-62 (1985).

298.   K. Zayat and Z. Bayasi, *ACI Mater. J.* 93(2), 178-181 (1996).

299.   P. Soroushian, F. Aouadi and M. Nagi, *ACI Mater. J.* 88(1), 11-18 (1991).

300.   B. Mobasher and C. Y. Li, *ACI Mater. J.* 93(3), 284-292 (1996).

301.   N. Banthia, A. Moncef, K. Chokri and J. Sheng, *Can. J. Civil Eng.* 21(6), 999-1011 (1994).

302.   B. Mobasher and C.Y. Li, *Infrastructure: New Materials and Methods of Repair,* Proc. Mater. Eng. Conf., n 804, (1994), ASCE, New York, NY, pp. 551-558.

303.   P. Soroushian, M. Nagi and J. Hsu, *ACI Mater. J.* 89(3), 267-276 (1992).

304.   P. Soroushian, *Construction Specifier* 43(12), 102-108 (1990).

305.   A.K. Lal, *Batiment Int./Building Research & Practice* 18(3), 153-161 (1990).

306.   S.B. Park, B.I. Lee and Y.S. Lim, *Cem. Concr. Res.* 21(4), 589-600 (1991).

307.   S.B. Park and B.I. Lee, *High Temperatures – High Pressures* 22(6), 663-670 (1990).

308.   P. Soroushian, M. Nagi and A. Okwuegbu, *ACI Mater. J.* 89(5), 491-494 (1992).

309.   M. Pigeon, M. Azzabi and R. Pleau, *Cem. Concr. Res.* 26(8), 1163-1170 (1996).

310.   N. Banthia, K. Chokri, Y. Ohama and S. Mindess, *Adv. Cem. Based Mater.* 1(3), 131-141 (1994).

311.   N. Banthia, C. Yan and K. Sakai, *Com. Concr. Composites* 20(5), 393-404 (1998).

312.   T. Urano, K. Murakami, Y. Mitsui and H. Sakai, *Composites – Part A: Applied Science & Manufacturing* 27(3), 183-187 (1996).

313.   A. Ali and R. Ambalavanan, *Indian Concrete J.* 72(12), 669-675.

314.   P. Chen, X. Fu and D.D.L. Chung, *Cem. Concr. Res.* 25(3), 491-496 (1995).

315.   M. Zhu and D.D.L. Chung, *Cem. Concr. Res.* 27(12), 1829-1839 (1997).

316.   M. Zhu, R. C. Wetherhold and D.D.L. Chung, *Cem. Concr. Res.,* 27(3), 437-451 (1997).

317.   P. Chen and D.D.L. Chung, *Smart Mater. Struct.* 2, 22-30 (1993).

318. P. Chen and D.D.L. Chung, *Composites*, Part B 27B, 11-23 (1996).

319. P. Chen and D.D.L. Chung, *J. Am. Ceram. Soc.* 78(3), 816-818 (1995).

320. D.D.L. Chung, *Smart Mater. Struct.* 4, 59-61 (1995).

321. P. Chen and D.D.L. Chung, *ACI Mater. J.* 93(4), 341-350 (1996).

322. X. Fu and D.D.L. Chung, *Cem. Concr. Res.* 26(1), 15-20 (1996).

323. X. Fu, E. Ma, D.D.L. Chung and W. A. Anderson, *Cem. Concr. Res.* 27(6), 845-852 (1997).

324. X. Fu and D.D.L. Chung, *Cem. Concr. Res.* 27(9), 1313-1318 (1997).

325. X. Fu, W. Lu and D.D.L. Chung, *Cem. Concr. Res.* 28(2), 183-187 (1998).

326. Z. Shi and D.D.L. Chung, *Cem. Concr. Res.* 29(3), 435-439 (1999).

327. Q. Mao, B. Zhao, D. Sheng and Z. Li, *J. Wuhan U. Tech.*, Mater. Sci. Ed. 11(3), 41-45 (1996).

328. Q. Mao, B. Zhao, D. Shen and Z. Li, *Fuhe Cailiao Xuebao/Acta Materiae Compositae Sinica*, 13(4), 8-11 (1996).

329. M. Sun, Q. Mao and Z. Li, *J. Wuhan U. Tech.*, Mater. Sci. Ed. 13(4), 58-61 (1998).

330. B. Zhao, Z. Li and D. Wu, *J. Wuhan Univ. Tech.*, Mater. Sci. Ed. 10(4), 52-56 (1995).

331. S. Wen and D.D.L. Chung, *Cem. Concr. Res.* 29(6), 961-965 (1999).

332. M. Sun, Z. Li, Q. Mao and D. Shen, *Cem. Concr. Res.* 28(4), 549-554 (1998).

333. M. Sun, Z. Li, Q. Mao and D. Shen, *Cem. Concr. Res.* 28(12), 1707-1712 (1998).

334. S. Wen and D.D.L. Chung, *Cem. Concr. Res.*, 29(12), 1989-1993 (1999).

335. D. Bontea, D.D.L. Chung and G.C. Lee, *Cem. Concr. Res.* 30(4), 651-659 (2000).

336. S. Wen and D.D.L. Chung, *Cem. Concr. Res.* 30(12), 1979-1982 (2000).

337. J. Lee and G. Batson, *Materials for the New Millennium*, Proc. 4th Mater. Eng. Conf., vol. 2, (1996), ASCE, New York, NY, pp. 887-896.

338. X. Fu and D.D.L. Chung, *ACI Mater. J.* 96(4), 455-461 (1999).

339. X. Fu and D.D.L. Chung, *Composite Interfaces* 6(2), 81-92 (1999).

340. J. Hou, X. Fu and D.D.L. Chung, *Cem. Concr. Res.* 27(5), 679-684 (1997).

341. S. Wen and D.D.L. Chung, *Cem. Concr. Res.* 30(2), 327-330 (2000).

342. X. Fu and D.D.L. Chung, *Cem. Concr. Res.* 25, 1397 (1995).

343. Y. Xu and D.D.L. Chung, *ACI Mater. J.* 97(3), 333-342 (2000).

344. Y. Xu and D.D.L. Chung, *Cem. Concr. Res.* 30(2), 241-245 (2000).

345. Y. Xu and D.D.L. Chung, *Cem. Concr. Res.* 30(7), 1175-1178 (2000).

346. Y. Xu and D.D.L. Chung, *Cem. Concr. Res.* 30(8), 1305-1311 (2000).

347. T.A. Bürge, Bond in Concrete, P. Bartos, Ed., Applied Science Publishers, London, 1982, p. 273-281.

348. O.E. Gjorv, P.J.M. Monteiro and P.K. Mehta, *ACI Mater. J.* 87(6), 573 (1990).

349. X. Fu and D.D.L. Chung, *Cem. Concr. Res.* 28(4), 487-492 (1998).

350. X. Fu and D.D.L. Chung, *Cem. Concr. Res.* 26(2), 189-194 (1996).

351. ACI Committee 548, *ACI Mater. J.* 91, 511 (1994).

352. B.S. Wyatt, *Corrosion Science* 35(5-8), pt 2, 1601-1615 (1993).

353. S.C. Das, *Structural Engineer* 71(22), 400-403 (1993).

354. J.S. Tinnea and R.P. Brown, *Materials for the New Millennium, Proc. 1996 4th Materials Engineering Conf.,* vol 2, (1996), ASCE, New York, NY, 1531-1539.

355. I. Solomon, M.F. Bird, and B. Phang, *Corrosion Science* 35(5-8), pt. 2, 1649-1660 (1993).

356. J. Hou and D.D.L. Chung, *Cem. Concr. Res.* 27(5), 649-656 (1997).

357. R.J. Kessler, R.G. Powers, and I.R. Lasa, *Materials Performance* 37(1), 12-15 (1998).

358. R.J. Brousseau and G.B. Pye, *ACI Mater. J.* 94(4), 306-310 (1997).

359. F. Papworth and R. Ratcliffe, *Concr. Int.* 16(10), 39-44 (1994).

360. K.E.W. Coulson, T.J. Barlo, and D.P. Werner, *Oil & Gas J.* 89(41), 80-84 (1991).

361. V. Dunlap, *Proc. Conf. Cathodic Protection of Reinforced Concrete Bridge Decks*, NACE, Houston, TX, pp. 131-136.

362. B. Heuze, *Materials Performance* 19(5), 24-33 (1980).

363. R. Pangrazzi, W.H. Hartt, and R. Kessler, *Corrosion* 50(3), 186-196 (1994).

364. R.E. Weyers, W. Pyc, and M.M. Sprinkel, *ACI Mater. J.* 95(5), 546-557. (1998)

365. S.W. Poon and I.F. Tasker, *Asia Engineer* 26(8), 17 (1998).

366. R.D. Lampton Jr. and D. Schemberger, *Materials for the New Millenium, Proc. 1996 4th Materials Engineering Conf.,* vol. 2, (1996), ASCE, New York, NY, 1209-1218.

367. J.L. Smith and Y.P. Virmani, *Public Roads* 60(2), 6-12 (1996).

368. B. Neffgen, *European Coatings J.* (10), 700-703 (1996).

369. J. Shubrook, *Plant Engineering* (Barrington, Illinois) 50(10), 99-100 (1996).

370. J.S. McHattie, I.L. Perez, and J.A. Kehr, *Cem. Concr. Composites* 18(2), 93-103 (1996).

371. J. Hartley, *Steel Times* 224(1), 23-24 (1996).

372. K. Thangavel, N.S. Rengaswamy, and K. Balakrishnan, *Indian Concrete J.* 69(5), 289-293 (1995).

373. L.K. Aggarwal, K.K. Asthana, and R. Lakhani, *Indian Concrete J.* 69(5), 269-273 (1995).

374. H.O. Hasan, J.A. Ramirez, and D.B. Cleary, *Better Roads* 65(5), 2 pp (1995).

375. K. Kahhaleh, J. Jirsa, R. Carrasquillo, and H. Wheat, *Infrastructute: New Materials and Methods of Repair, Proc. 3rd Materials Engineering Conf.*, no. 804, (1994), ASCE, New York, NY, 8-15.

376. R.G. Mathey and J.R. Clifton, *Proc. Struct. Congr. 94*, (1994), ASCE, New York, NY, pp. 109-115.

377. P. Schiessl and C. Reuter, *Beton – und Stahlbetonbau*, 87(7), 171-176 (1992).

378. R. Korman, *ENR* (Engineering News-Record) 228(19), 9 (1992).

379. K.C. Clear, *Concrete International: Design & Construction* 14(5), 58, 60-62 (1992).

380. T.E. Cousins, D.W. Johnston, and P. Zia, *ACI Mater. J.* 87(4), 309-318 (1990).

381. K.W.J. Treadaway and H. Davies, *Structural Engineer* 67(6), 99-108 (1989).

382. R.A. Treece and J.O. Jirsa, *ACI Mater. J.* 86(2), 167-174 (1989).

383. S. Muthukrishnan and S. Guruviah, *Transactions of the SAEST* (Society for Advancement of Electrochemical Science & Technology) 23(2-3), 183-188 (1988).

384. T.D. Lin, R.I. Zwiers, S.T. Shirley, and R.G. Burg, *ACI Mater. J.* 85(6), 544-550 (1988).

385. H.A. El-Sayed, F.H. Mosalamy, A.F. Galal, and B.A. Sabrah, *Corrosion Prevention & Control* 35(4), 87-92 (1988).

386. D.P. Gustafson, *Civil Engineering* (New York) 58(10), 38-41 (1988).

387. L. Salparanta, *Valt Tek Tutkimuskeskus Tutkimuksia* 521, 48 p (1988).

388. R.J. Higgins, *Concrete Plant & Production* 5(6), 197-198 (1987).

389. R.J. Higgins, *Concrete Plant & Production* 5(4), 131-132 (1987).

390. H.A. El-Sayed, M.M. Kamal, S.N. El-Ebiary, and H. Shahin, *Corrosion Prevention & Control* 34(1), 18-23 (1987).

391. B.W. McLean and A.J.R. Bridges, *U.K. Corrosion '85*, Corrosion and Preparation, Corrosion Monitoring, Materials Selection, Inst. Of Corrosion Science & Technology, Vol 2, (1985), Birmingham, Engl., pp. 11-17.

392. S.L. Lopata, *ASTM Special Technical Publication 841*, New Concepts for Coating Protection of Steel Structures, ASTM, Philadelphia, PA, pp. 5-9.

393. T. Arai, K. Shirakawa, N. Mikami, S. Koyama, and A. Yamazaki, *Sumitomo Metals* 36(3), 53-71 (1984).

394. J. Clifton, *To Build and Take Care of What We Have Built with Limited Resources*, CIB 83, 9th CIB Congress, News Review, Natl. Swedish Inst. For Building Research, (1983), Gavle, Swed., pp. 68-69.

395.   K. Kobayashi and K. Takewaka, *International J. Cement Composites & Lightweight Concrete,* 6(2), 99-116 (1984).
396.   V.R. Subramanian, *Indian Concrete J.* 70(7), 383-385 (1996).
397.   S.R. Yeomans, *Hong Kong Institution of Engineers Transactions* 2(2), 17-28 (1995).
398.   N. Gowripalan and H.M. Mohamed, *Cem. Concr. Res.* 28(8), 1119-1131 (1998).
399.   O.A. Kayyali and S.R. Yeomans, *Construction & Building Materials,* 9(4), 219-226 (1995).
400.   S.R. Yeomans, *Corrosion,* 50(1), 72-81 (1994).
401.   F.H. Rasheeduzzafar, F.H. Dakhil, M.A. Bader, and M.M. Khan, *ACI Mater. J.* 89(5), 439-448 (1992).
402.   N.S. Rengaswamy, S. Srinivasan, and T.M. Balasubramanian, *Transactions of the SAEST* (Society for Advancement of Electrochemical Science & Technology) 23(2-3), 163-173 (1988).
403.   D.B. McDonald, Y.P. Virmani, and D.F. Pfeifer, *Concrete International* 18(11), 39-43 (1996).
404.   H. Kishikawa, H. Miyuki, S. Hara, M. Kamiya, and M. Yamashita, *Sumitomo Search* (60), 20-26 (1998).
405.   J. Hou, X. Fu, and D.D.L. Chung, *Cem. Concr. Res.* 27(5), 679-684 (1997).
406.   N.S. Rengaswamy, R. Vedalakshmi, and K. Balakrishnan, *Corrosion Prevention & Control* 42(6), 145-150 (1995).
407.   J.M. Gaidis and A.M. Rosenberg, *Cem., Concr. & Aggregates* 9(1), 30-33 (1987).
408.   N.S. Berke, *Concr. Int.: Design & Construction* 13(7), 24-27 (1991).
409.   C.K. Nmai, S.A. Farrington, and G.S. Bobrowski, *Concr. Int.: Design & Construction* 14(4), 45-51 (1992).
410.   N.S. Rengaswamy, V. Saraswathy, and K. Balakrishnan, *J. Ferrocement* 22(4), 359-371 (1992).
411.   C.K. Nmai, M.A. Bury, and H. Farzam, *Concr. Int.: Design & Construction* 16(4), 22-25 (1994).
412.   C.K. Nmai, M.A. Bury and H. Farzam, *Concr. Int.* 16(4), 22-25 (1994).
413.   S.H. Tantawi, *J. Mater. Sci. & Technology* 12(2), 95-99 (1996).
414.   R.J. Scancella, *Materials for the New Millenium,* Proc. 1996 4th Materials Engineering Conf., vol 2, (1996), ASCE, New York, NY, pp. 1276-1280.
415.   J.R. Miller and D.J. Fielding, *Concr. Int.,* 19(4), 29-34 (1997).
416.   I.Z. Selim, *J. Mater. Sci. & Technology* 14(4), 339-343 (1998).
417.   D. Bjegovic and B. Miksic, *Mater. Performance* 38(11), 52-56 (1999).
418.   M. Tullmin, L. Mammoliti, R. Sohdi, C.M. Hansson and B.B. Hope, *Cem., Concr. & Aggregates* 17(2), 134-144 (1995).

419. C.M. Hansson, L. Mammoliti and B.B. Hope, *Cem. Concr. Res.* 28(12), 1775-1781 (1998).

420. N.S. Berke, M.P. Dallaire, R.E. Weyers, M. Henry, J.E. Peterson and B. Prowell, *ASTM Special Technical Publication,* Symposium on Corrosion Forms and Control for Infrastructure, no. 1137, (1991), ASTM, Philadelphia, PA, pp. 300-327.

421. N.S. Berke and A. Rosenberg, *Transportation Research Record* (1211), 18-27 (1989).

422. N.S. Berke, *Transportation Research Record* (1204), 21-26 (1988).

423. J. Hou and D.D.L. Chung, *Corrosion Sci.* 42(9), 1489-1507 (2000).

424. J.G. Cabrera and P.A. Claisse, *Construction & Building Mater.* 13(7), 405-414 (1999).

425. O.E. Gjorv, *ACI Mater. J.* 92(6), 591-598 (1995).

426. J.G. Cabrera, P.A. Claisse and D.N. Hunt, *Construction & Building Mater.* 9(2), 105-113 (1995).

427. N.R. Jarrah, O.S.B. Al-Amoudi, M. Maslehuddin, O.A. Ashiru and A.I. Al-Mana, *Construction & Building Mater.* 9(2), 97-103 (1995).

428. T. Lorentz and C. French, *ACI Mater. J.* 92(2), 181-190 (1995).

429. S.A. Khedr and A.F. Idriss, *J. Materials in Civil Engineering* 7(2), 102-107 (1995).

430. C. Ozyildirim, *ACI Mater. J.* 91(2), 197-202 (1994).

431. J.T. Wolsiefer Sr., *Utilization of Industrial By-Products for Construction Materials, Proc. ASCE National Convention and Exposition,* (1993), ASCE, New York, NY, pp. 15-29.

432. K. Torii, M. Kawamura, T. Asano, and M. Mihara, *Zairyo/Journal of the Society of Materials Science,* Japan, vol 40, no. 456, (1991), pp. 1164-1170.

433. H.T. Cao and V. Sirivivantnanon, *Cem. Concr. Res.* 21(2-3), 316-324 (1991).

434. Anon., *Concr. Construction* 33(2), 6 p (1988).

435. K.P. Fischer, O. Bryhn, and P. Aagaard, *Publikasjon – Norges Geotekniske Institutt,* (161), 9 p (1986).

436. O. Gautefall and O. Vennesland, *Nordic Concrete Research* 2, 17-28 (1983).

437. P.J.M. Monteiro, O.E. Gjorv and P.K. Mehta, *Cem. Concr. Res.* 15(5), 781-784 (1985).

438. O. Vennesland and O.E. Gjorv, *Publication SP – American Concrete Institute 79,* Fly Ash, Silica Fume, Slag & Other Mineral By-Products in Concrete, vol 2, (1979), American Concrete Inst., Detroit, MI, pp. 719-729.

439. A.H. Ali, *Corrosion Prevention & Control* 46(3), 76-81 (1999).

440. A.H. Ali, B. El-Sabbagh, and H.M. Hassan, *Corrosion Prevention & Control* 45(6), 173-180 (1998).

441.   M. Maslehuddin, H. Saricimen, and A.I. Al-Mana, *ACI Mater. J.* 84(1), 42-50 (1987).
442.   M. Maslehuddin, A.I. Al-Mana, M. Shamim, and H. Saricimen, *ACI Mater. J.* 86(1), 58-62 (1989).
443.   O.S.B. Al-Amoudi, M. Maslehuddin, and I.M. Asi, *Cem., Concr. & Aggregates* 18(2), 71-77 (1996).
444.   M. Maslehuddin, A.I. Al-Mana, H. Saricimen, and M. Shamim, *Cem., Concr. & Aggregates* 12(1), 24-31 (1990).
445.   K. Babaei and N.M. Hawkins, *ASTM Special Technical Publication,* Symposium on Corrosion Forms and Control for Infrastructure, no. 1137, (1991), ASTM, Philadelphia, PA, 140-154.
446.   S.H. Okba, A.S. El-Dieb, and M.M. Reda, *Cem. Concr. Res.,* 27(6), 861-868 (1997).
447.   S.X. Wang, W.W. Lin, S.A. Ceng, and J.Q. Zhang, *Cem. Concr. Res.* 28(5), 649-653 (1998).
448.   F. Andrews-Phaedonos, *Transport Proc. 1998 19th ARRB Conference of the Australian Road Research Board,* (1998), ARRB Transport Research Ltd., Vermont, Australia, pp. 245-262.
449.   J.-Z. Zhang, I.M. McLoughlin, and N.R. Buenfeld, *Cem. & Concr. Compos.* 20(4), 253-261 (1998).
450.   R.N. Swamy and S. Tanikawa, *Materials & Structures* 26(162), 465-478 (1993).
451.   B.G. Callaghan, *Corrosion Sci.* 35(5-8), pt 2, 1535-1541 (1993).
452.   K. Kumar, R. Vedalakshmi, S. Pitchumani, A. Madhavamayandi and N.S. Rengaswamy, *Indian Concr. J.* 70(7), 359-364 (1996).
453.   K. Murphy, S. Zhang, and V.M. Karbhari, *Int. SAMPE Symp. & Exhib.* 44(II), 2222-2230 (1999).
454.   H. Saadatmanesh and F. Tannous, *Proc. 1997 Int. Seminar on Repair and Rehabilitation of Reinforced Concrete Structures: The State of the Art,* (1998), ASCE, Reston, VA, pp. 120-133.
455.   S.S. Faza, *Materials for the New Millennium, Proc. 1996 4th Materials Engineering Conf.,* vol 2, (1996), ASCE, New York, NY, pp. 905-913.
456.   S. Loud, *SAMPE J.* 32(1), 5 p (1996).
457.   T.R. Gentry and M. Husain, *J. Compos. for Construction* 3(2), 82-86 (1999).
458.   H.C. Boyle and V.M. Karbhari, *Polym.-Plastics Tech. & Eng.* 34(5), 697-720 (1995).

2

Introduction to Multifunctional Cement-Based Materials

2.1   Structural applications

Structural applications refer to applications that require mechanical performance (e.g., strength, stiffness and vibration damping ability) in the material, which may or may not bear the load in the structure. In case the material bears the load, the mechanical property requirements are particularly stringent. An example is a building in which steel-reinforced concrete columns bear the load of the structure and unreinforced concrete architectural panels cover the face of the building. Both the columns and the panels serve structural applications and are structural materials, though only the columns bear the load of the structure. Mechanical strength and stiffness are required of the panels, but the requirements are more stringent for the columns.

Structures include buildings, bridges, piers, highways, landfill cover, aircraft, automobiles (body, bumper, shaft, window, engine components, brake etc.), bicycles, wheelchairs, ships, submarines, machinery, satellites, missiles, tennis rackets, fishing rods, skis, pressure vessels, cargo containers, furniture, pipelines, utility poles, armor, utensils, fasteners, etc.

In addition to mechanical properties, a structural material may be required to have other properties, such as low density (lightweight) for fuel saving in the case of aircraft and automobiles, for high speed in the case of race bicycles, and for handleability in the case of wheelchairs and armor. Another property that is often required is corrosion resistance, which is desirable for the durability of all structures, particularly automobiles and bridges. Yet another property that may be required is the ability to withstand high temperatures and/or thermal cycling, as heat may be encountered by the structure during operation, maintenance or repair.

A relatively new trend is for a structural material to be able to serve functions other than the structural function, so that the material becomes multifunctional (i.e., killing two or more birds with one stone, thereby saving cost and simplifying design). An example of a nonstructural function is the sensing of damage (Chapter 4). Such sensing, also called structural health monitoring, is valuable for the prevention of hazards. It is particularly important to aging aircraft and bridges. The sensing function can be attained by embedding sensors (such as optical fibers, the damage or strain of which affects the light throughput) in the structure. However, the embedding usually causes degradation of the mechanical properties, and the embedded devices are costly and poor in durability compared to the structural material. Another way to attain the sensing function is to detect the change in property (e.g., the electrical resistivity) of the structural material due to damage. In this way, the structural material serves as its own sensor and is said to be self-sensing.

Mechanical performance is basic to the selection of a structural material. Desirable properties are high strength, high modulus (stiffness), high ductility, high toughness (energy absorbed in fracture) and high capacity for vibration damping. Strength, modulus and ductility can be measured under tension, compression or flexure at various loading rates, as indicated by the type of loading on the structure. A high compressive strength does not imply a high tensile strength. Brittle materials tend to be stronger under compression than tension due to the microcracks in them. High modulus does not imply high strength, as the modulus describes the elastic deformation behavior whereas the strength describes the fracture behavior. Low toughness does not imply a low capacity for vibration damping, as the damping (energy dissipation) may be due to slipping at interfaces in the material, rather than being due to the shear of a viscoelastic phase in the material. Other desirable mechanical properties are fatigue resistance, creep resistance, wear resistance and scratch resistance.

Structural materials are predominantly metal-based, cement-based and polymer-based materials, although they also include carbon-based and ceramic-based materials, which are valuable for high temperature structures. Among the metal-based structural materials, steel and aluminum alloys are dominant. Steel is advantageous in the high strength, whereas aluminum is advantageous in the low density. For high temperature applications, intermetallic compounds (such as NiAl) have emerged, though they suffer from their brittleness. Metal-matrix composites are superior to the corresponding metal matrix in the high modulus, high creep resistance and low thermal expansion coefficient, but they are expensive due to the processing cost.

Among the cement-based structural materials, concrete is dominant. Although concrete is an old material, improvement in the long-term durability is needed, as suggested by the degradation of bridges and highways all over the U.S. The improvement pertains to decrease in the drying shrinkage (shrinkage of the concrete during curing or hydration), as the shrinkage can cause cracks. It also pertains to decrease in the fluid permeability, as water permeating into steel reinforced concrete can cause corrosion of the reinforcing steel. Moreover, it pertains to improvement in the freeze-thaw durability, which is the ability of the concrete to withstand temperature variations between temperatures below 0°C (freezing of water in concrete) and those above 0°C (thawing of water in concrete).

Among the polymer-based structural materials, fiber-reinforced polymers are dominant, due to their combination of high strength and low density. All polymer-based materials suffer from their inability to withstand high temperatures. This inability can be due to the degradation of the polymer itself or, in the case of a polymer-matrix composite, due to the thermal stress resulting from the thermal expansion mismatch between the polymer matrix and the fibers. (The coefficient of thermal expansion is typically much lower for the fibers than for the matrix.)

Most structures involve joints, which may be attained by welding, brazing, soldering, the use of adhesives, or fastening. The structural integrity of joints is critical to the integrity of the overall structure.

As structures can degrade or be damaged, repair may be needed. Repair often involves the use of a repair material, which may be the same as or different from the original material. For example, a damaged concrete column may be repaired by removing the damaged portion and patching with a fresh concrete mix. A superior but much more costly way involves the abovementioned patching, followed by wrapping the column with continuous carbon or glass fibers, using epoxy as the adhesive between the fibers and the column. Due to the tendency for the molecules of a thermoplastic polymer to move upon heating, joining of two thermoplastic parts can be attained by liquid state or solid state welding, thereby facilitating repair of a thermoplastic structure. In contrast, the molecules of a thermosetting polymer do not tend to move, and repair of a thermoset structure needs to involve other methods, such as the use of adhesives.

Corrosion resistance is desirable for all structures. Metals, due to their electrical conductivity, are particularly prone to corrosion. In contrast, polymers and ceramics, due to their poor conductivity, are much less prone to corrosion. Techniques of corrosion protection include the use of a sacrificial anode (i.e., a material that is more active than the material to be protected, so that it is the party that corrodes) and cathodic protection (i.e., the application of a voltage which causes electrons to go into the material to be protected, thereby making the material a cathode). The implementation of the first technique simply involves attaching the sacrificial anode material to the material to be protected. The implementation of the second technique involves applying an electrical contact material on the surface of the material to be protected and passing an electric current through wires embedded in the electrical contact. The electrical contact material must be a good electrical conductor and be able to adhere to the material to be protected, in addition to being wear-resistant and scratch-resistant.

Vibration damping (Chapter 8) is desirable for most structures. It is commonly attained by attaching to or embedding in the structure a viscoelastic material, such as rubber. Upon vibration, shear deformation of the viscoelastic material causes energy dissipation. However, due to the low strength and modulus of the viscoelastic material compared to the structural material, the presence of the viscoelastic material (especially if it is embedded) lowers the strength and modulus of the structure. A more ideal way to attain vibration damping is to modify the structural material itself, so that it maintains its high strength and modulus while providing damping. In the case of a composite material being the structural material, the modification can involve the addition of a filler (particles or fibers) with a very small size, so that the total filler-matrix interface area is large and slippage at the interface during vibration provides a mechanism of energy dissipation.

2.2   Multifunctionality and electrical conduction behavior

Concrete is a common structural material. Whether it is load bearing or not in a structure, its strength and stiffness are important. Although purely structural applications dominate, combined structural and nonstructural applications are increasingly important as smart structures and electronics become more common. Such combined applications are facilitated by multifunctional concretes, which exhibit good structural properties as well as some desirable nonstructural properties. The nonstructural properties can be vibration damping ability, strain sensing ability, electromagnetic/magnetic shielding ability, electrical conductivity (for grounding, lightning protection, electrical contacts and cathodic protection) and thermal insulation ability. As a multifunctional concrete serves two or more functions, its use diminishes or eliminates the need for conventional nonstructural materials, such as elastomers for vibration damping, strain gages for strain sensing, metal wire mesh for electromagnetic/magnetic shielding, metals for electrical conduction and foams for thermal insulation. As these conventional nonstructural materials are usually much more expensive than concrete, diminishing or eliminating the need for them reduces cost. Furthermore, these conventional nonstructural materials are usually less durable than concrete, so diminishing or eliminating the need for them improves the durability. In addition, diminishing or eliminating the need for them simplifies the design, reduces the assembly cost and allows the structure to be more compact.

As structural materials, cement-based materials have received much attention in terms of the mechanical properties, but relatively little attention in terms of the electrical conduction properties. Nevertheless, the electrical properties are relevant to the use of the structural materials for nonstructural functions, such as sensing and electromagnetic interference (EMI) shielding. The ability to provide nonstructural functions allows a structural material to be multifunctional, thus saving cost and enhancing durability. Furthermore, the electrical conduction properties shed light on the structure of the materials, particularly concerning the interfaces in the composite materials. Therefore, for both technological and fundamental reasons, the electrical conduction properties of cement-matrix composites are of interest.

2.3 Background on electrical conduction behavior

2.3.1 Charge transport

In order for a material to conduct electricity, electrical charges (or charged particles) must be available to move through the material. The movement of charged particles constitutes the electric current. Charged particles which serve to conduct electricity are commonly known as charge carriers, which can be positively or negatively charged.

Mobile charged particles are commonly electrons, which are very light in mass and so can move in a solid quite easily. Metals have valence electrons which are weakly bound to the nuclei. Therefore, these electrons, called free electrons, can move easily in a metal, making the metal have a high electrical conductivity (or a low electrical resistivity). This is why metals are in general good electrical conductors. On the other hand, covalent solids (e.g., diamond, pure silicon) have negligible free electrons because the valence electrons in a covalent solid are localized between covalently bonded atoms so that they are not free to move away from their parent atoms. Ionic solids also have negligible free electrons because the electron transfer causes energetically favorable electronic configurations in the cations and anions, so that the valence electrons are all tightly bound to the ions. Electrical conduction in ionic solids is achieved through the movement of the ions, which are charged particles. However, due to the relatively high mass of ions compared with electrons, ions cannot move very easily in the solid. Therefore, ionic and covalent solids are usually poor electrical conductors (i.e., electrical insulators).

The flow of electrical charge is known as the current (I). This means that I is the rate of charge passage, or the amount of charge passing through per unit time. Hence,

$$I = \frac{charge}{time}. \tag{2.1}$$

The unit of charge is the coulomb and that of I is the ampere, where

$$1 \text{ ampere} = \frac{1 \text{ coulomb}}{sec}. \tag{2.2}$$

Rather than talking in terms of the current (I), it is sometimes more convenient to talk in terms of the current density $(\tilde{J})$, which is the current per unit area:

$$\tilde{J} = \frac{I}{A}, \tag{2.3}$$

where A is the specimen cross-sectional area.

The flow of electrical charge constitutes electricity. What makes an electrical charge or an electron flow through a solid in a certain direction?

There are two causes of such a flow – a voltage gradient or a charge carrier concentration gradient. A voltage gradient means the presence of an electric field; it can be provided by, say, putting a power source across the conducting material. A charge carrier concentration gradient means non-uniformity in the charge carrier distribution in the conducting material.

2.3.1.1 Transport due to voltage gradient

When placed between a positive electrode and a negative electrode, positive charge carriers are attracted toward the negative electrode while negative charge carriers are attracted toward the positive electrode, as illustrated in Fig. 2.1. This is a consequence of Coulomb's Law. Since the positive electrode is at a higher voltage than the negative electrode, the two electrodes provide a voltage gradient or an electric field. As shown in Fig. 2.1, the positive charge carriers move down the voltage gradient while the negative charge carriers move up the voltage gradient.

When a power source such as a battery is connected to a conductor, as illustrated in Fig. 2.2, a current results. The current through the specimen may be measured with a voltmeter. These quantities are related through Ohm's Law:

$$V = I R , \qquad (2.4)$$

where V is the voltage drop in volts across the specimen, R is the electrical resistance of the specimen in ohms (Ω), and I is the current going through the circuit. The variation of V with I is linear, as shown in Fig. 2.3; the slope of the line is R.

For a given material, the resistance R varies with simple dimensions, being directly proportional to the specimen length ℓ and inversely proportional to the specimen cross-sectional area A (Fig. 2.4). Hence,

$$R \propto \frac{\ell}{A} .$$

Let the constant of proportionality be $1/\sigma$, which is called the electrical resistivity. Hence,

$$R = \frac{1}{\sigma} \frac{\ell}{A} . \qquad (2.5)$$

The electrical resistivity is a property of the specimen material. Since the units of R, ℓ and A are Ω, m and m^2 respectively, the unit of $1/\sigma$ is Ω.m. For example, the electrical resistivity of silicon at 300 K is 2300 Ω.m.

The reciprocal of $1/\sigma$ is σ, which is called the electrical conductivity. The unit of σ is Ω^{-1}.m^{-1}. Because of Eq. (2.4),

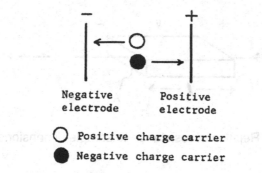

Fig. 2.1   Response of charge carriers to a voltage gradient.

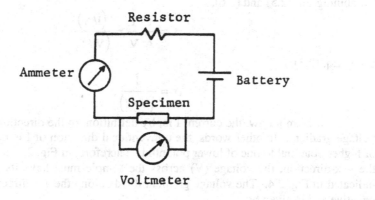

Fig. 2.2   A schematic circuit.

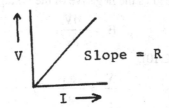

Fig. 2.3   Ohm's Law.

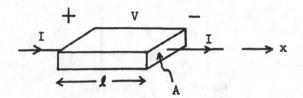

Fig. 2.4   Relation of resistance to sample dimensions.

$$R = \frac{V}{I}. \tag{2.6}$$

Combining Eq. (2.5) and (2.6),

$$\sigma = \frac{\ell}{A} \frac{I}{V} = \frac{(I/A)}{(V/\ell)}. \tag{2.7}$$

Using Eq. (2.3),

$$\sigma = \frac{\tilde{J}}{(V/\ell)}. \tag{2.8}$$

In Ohm's Law, the current I is, by definition, in the direction down the voltage gradient. In other words, the conventional direction of I is from a point of higher potential to one of lower potential. Therefore, in Fig. 2.4, for I to be in the + x-direction, the voltage (V) across the sample must have its polarity as indicated in Fig. 2.4. The voltage gradient dV/dx along the + x direction is thus negative and is given by

$$\frac{dV}{dx} = -\frac{V}{\ell}. \tag{2.9}$$

The electric field E is defined as the negative of the voltage gradient, i.e.,

$$E = -\frac{dV}{dx}. \tag{2.10}$$

Combining Eq. (2.9) and (2.10),

$$\frac{V}{\ell} = -\frac{dV}{dx} = E.$$

Hence, Eq. (2.8) can be written as

$$\sigma = \frac{\tilde{J}}{E}. \tag{2.11}$$

Eq. (2.11) can be considered an alternative way of expressing Ohm's law, since σ is related to R, J is related to I, and E is related to V.

Due to thermal energy, charge carriers are constantly moving, even in the absence of a voltage gradient or a charge carrier concentration gradient. This is particularly true for electrons, which are extremely small and light. In fact, the constant motion of electrons is required by the Uncertainty Principle, which says that the position and velocity of an electron cannot be simultaneously known. This is why we think of an electron as an electron cloud. However, the motion is random, so that there is no net velocity in any direction. In the presence of a voltage gradient, the charge carriers preferentially move in a certain direction, so that a nonzero net speed results. This flow is known as drift, and the net speed is called the drift speed (v), which is proportional to the electric field (E) and depends on the solid in which the charge carriers move.

$$v \propto E$$

The constant of proportionality is μ, which is known as the mobility of the charge carriers.

$$v = \mu E. \tag{2.12}$$

The mobility depends on the charge carriers as well as the material in which the charge carriers move. Since the unit of v is $m.s^{-1}$ and that of E is $V.m^{-1}$, the unit of μ is $m.s^{-1}/V.m^{-1} = m^2/V.s$. The mobility of electrons in silicon at 300 K is 0.13 $m^2/V.s$.

The conductivity of a material is related to the drift speed or the mobility. Consider the flow of charge carriers of charge q per carrier in a wire of cross-sectional area A. Let there be an electric field E, which gives rise to a drift speed v for the carriers. The current I through the wire is, by definition, the charge passing through a certain cross section of the wire per unit time. Consider arbitrarily the shaded cross section shown in Fig. 2.5. Suppose that the electric field is such that the drift velocity of the charge carriers is in the direction from left to right, as shown in Fig. 2.5. In unit time (i.e., 1 s), each charge carrier moves by a distance v in this direction. Therefore, in unit time, all the charge carriers within a distance of v to the left of the shaded cross section will pass through this cross section. The volume of this length of the wire is vA. Let the number of charge carriers per unit volume (called the carrier concentration) be n in the material of the wire. Then the number of charge carriers in this volume of the wire is nvA. Since the charge of each carrier is q, the amount of charge passing through the shaded cross section of the wire per unit time is qnvA. Therefore,

$$I = qnvA.$$

The current density is thus

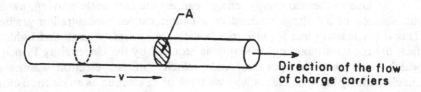

Direction of the flow
of charge carriers

Fig. 2.5   Current through a wire of cross-sectional area A.

$$\tilde{J} = \frac{I}{A} = \frac{qnvA}{A} = qnv \qquad (2.13)$$

Using Eq. (2.11),

$$\sigma = \frac{\tilde{J}}{E} = \frac{qnv}{E}. \qquad (2.14)$$

According to Eq. (2.12),

$$\frac{v}{E} = \mu.$$

Thus, Eq. (2.14) can be written as

$$\sigma = q n \mu. \qquad (2.15)$$

Eq. (2.15) means that the electrical conductivity of a material depends on the carrier concentration and the mobility of the carriers. Because the defects in a material interfere with the motion of the charge carriers, the mobility of the carriers decreases with increasing defect concentration. This is why cold work such as hammering or bending increases the electrical resistivity of a solid.

2.3.1.2 Transport due to charge carrier concentration gradient

Thermal energy causes the electrons in a solid to be in constant motion, but the motion is random unless (1) the electrons are in the presence of an electric field (which gives rise to drift) or (2) the electron concentration is not uniform in the solid (which gives rise to diffusion). The current arising from the diffusion of charged particles is called the diffusion current, which occurs only when there is a concentration gradient of the charged particles. In contrast, the current due to an electric field is called the conduction current or the drift current.

Consider the diffusion of negative charge carriers in a solid with a negative charge carrier concentration gradient dn/dx. The flux (J_n) of these negative carriers due to diffusion is given by

$$J_n = -D_n \frac{dn}{dx}, \qquad (2.16)$$

where D_n is the diffusion coefficient of the negative carriers. Let the charge on a carrier be $-q$ (where q is positive). Since J_n is the number of carriers flowing through unit area per unit time, the current density $\tilde{J}_n$ is given by

$$\tilde{J}_n = (-q) J_n. \qquad (2.17)$$

Substituting Eq. (2.17) in Eq. (2.16),

$$\tilde{J}_n = (-q)\left(-D_n \frac{dn}{dx}\right)$$

$$= q\, D_n \frac{dn}{dx}.$$

The minus sign in Eq. (2.17) makes the direction of the current density opposite to that of the negative carrier flow.

For the diffusion of positive charge carriers in a solid with a positive charge carrier concentration gradient dp/dx, the flux (J_p) of positive carriers is given by

$$J_p = -D_p \frac{dp}{dx}, \qquad (2.18)$$

where D_p is the diffusion coefficient of positive carriers. Let the charge on a positive carrier be q. The current density $\tilde{J}_p$ is given by

$$\tilde{J}_p = q\,J_p = q\left(-D_p \frac{dp}{dx}\right) = -q\,D_p \frac{dp}{dx}. \qquad (2.19)$$

The use of q instead of $-q$ in Eq. (2.19) makes the direction of the current density the same as that of the positive carrier flow.

2.3.1.3　Einstein relationship

Drift due to a voltage gradient and diffusion due to a concentration gradient are both manifestations of the random thermal motion of the carriers. Therefore, the mobility μ and the diffusion coefficient D are not independent. They are related as follows:

$$\frac{D_n}{\mu_n} = \frac{D_p}{\mu_p} = \frac{kT}{q} \qquad (2.20)$$

where k = Boltzmann's constant,
　　　　q = the magnitude of the charge on an electron
and　　T = temperature in K.

Eq. (2.20), known as the Einstein Relationship, means that D is proportional to μ, such that the proportionality constant is kT/q, which has the dimension of voltage and is called the thermal voltage or the volt-equivalent of temperature (V_T). Near room temperature (290 K), the thermal voltage kT/q = 25 mV.

2.3.2   Electronic energy bands

The electronic energy levels of an isolated (free) atom are discrete. In order of increasing energy, the electronic energy levels follow the trend 1s, 2s, 2p, 3s, 3p, etc. Consider, for example, sodium or Na, of which the electronic configuration is $1s^2\, 2s^2\, 2p^6\, 3s^1$. The energy levels of Na are shown in Fig. 2.6. An s subshell consists of just one orbital. Due to the Pauli exclusion principle, one orbital can only obtain up to two electrons. Therefore, an s subshell can contain a maximum of two electrons. A p subshell consists of three orbitals, so it can contain a maximum of six electrons.

Now consider a solid which consists of many atoms, say N atoms. Modern physics (quantum mechanics) shows that the atoms in a solid interact with one another such that the energy levels of all the atoms together are spread out over a range of energy. Thus, in contrast to isolated atoms which exhibit very narrow electronic energy levels, a solid exhibits electronic energy bands, as illustrated schematically in Fig. 2.7 for the case of Na. Since there are N atoms in the solid, an s band can contain a maximum of 2N electrons and a p band can contain a maximum of 6N electrons. The 1s, 2s and 2p bands are thus completely filled in solid Na. However, the 3s band is only half filled because it contains only N electrons while it can contain a maximum of 2N electrons. This is a consequence of the fact that the 3s subshell of each Na atom is half filled.

Quantum mechanics shows that the width of an energy band depends on how tightly bound the electrons are to the nuclei. The more tightly bound the electrons are in the band, the narrower the energy band. Since inner shell electrons are more tightly bound than the outer shell electrons, the energy bands become narrower as the principal quantum number decreases, as shown in Fig. 2.7. The widths of energy bands for outer shell electrons can be so large that adjacent

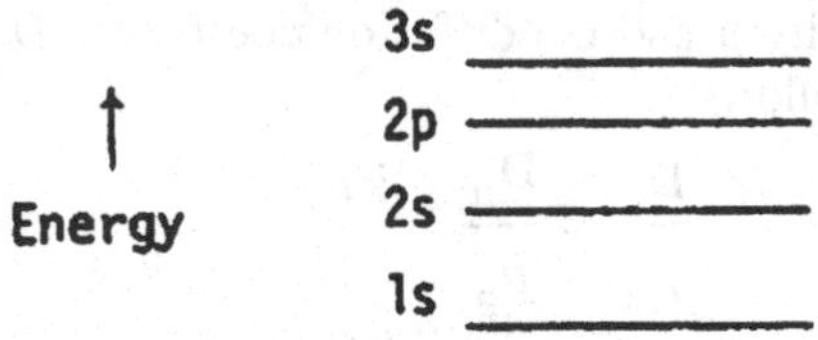

Fig. 2.6   Electronic energy levels of an isolated Na atom.

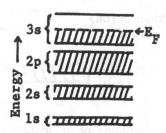

Fig. 2.7   Electronic energy bands of solid Na. E_F = Fermi energy.
Shaded energy ranges are occupied by electrons.

energy bands overlap. In fact, the band edges are not always sharp. For energy
bands that do not overlap, a gap of energy exists between adjacent bands. This gap
is known as the energy band gap (E_g), which is a range of energy that electrons are
forbidden to have. This gap has the same origin as the forbidden range of energy
between the discrete energy levels of an isolated atom.

For an electron to move so as to conduct electricity, it must be freed from
its parent atom and, moreover, possess kinetic energy. This means that the
electron, after being freed from its parent atom, must increase its energy, i.e., it
must move to a higher energy state, such that the higher energy state is not in the
forbidden energy gap and is not occupied by another electron. In the case of Na,
the 1s, 2s and 2p bands are all completely filled, whereas the 3s band is half filled.
In order for the 1s, 2s and 2p electrons to move to a higher energy state, they must
gain enough energy so as to be excited to the empty states in the 3s band. Since this
involves a considerable increase in energy, it is not likely for the electrons in the 1s,
2s and 2p bands to contribute to electrical conduction. On the other hand, the
electrons in the 3s band can be excited to the unoccupied states in the same band,
without crossing an energy band gap. Thus, the 3s electrons can move in the
presence of an electric field and contribute to electrical conduction. Therefore, in
the case of Na, the charge carriers are the 3s electrons, of which there is one per Na
atom. The carrier concentration in Na is thus equal to the number of Na atoms per
unit volume.

The Fermi energy (E_F) is defined as the energy below which all the
electron energy states are filled at 0 K. In the case of solid Na, the Fermi energy is
at the center of the 3s band, since only the lower half of the band is filled. The
Fermi energy of Na is indicated in Fig. 2.7.

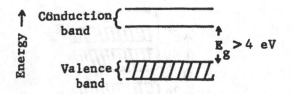

Fig. 2.8   Energy bands of an insulator. Shaded energy ranges
                   are occupied by electrons.

2.3.3   Metals, insulators, composites and semiconductors

2.3.3.1   Metals

Metals can accurately be defined as solids of which the valence electronic energy band is not completely filled. This means that the Fermi energy of a metal lies within the valence electronic energy band. Since the valence electrons can be excited without crossing an energy band gap, metals are characterized by high electrical conductivities. Because of overlap between adjacent valence bands, all of the valence electrons serve as charge carriers, so the larger the number of valence electrons per atom, the higher the carrier concentration in the solid. For example, Al has 3 valence electrons per atom, with the 3s and 3p bands overlapping, so it has a higher carrier concentration than Na, which has 1 valence electron per atom. Note that this definition of a metal agrees with the fact that atoms in a metal are held together by metallic bonding, which is provided by delocalized valence electrons. The delocalization of the valence electrons (free electrons) is necessary for these electrons to move large distances away from the parent nuclei.

2.3.3.2   Insulators

Whereas metals (good electrical conductors) have incompletely filled valence bands, insulators (poor electrical conductors) have completely filled valence bands. Furthermore, the energy band gap between the top of the filled valence band and the bottom of the empty higher energy band is large, typically greater than 4 eV. These energy bands are illustrated in Fig. 2.8. In order for the valence electrons to be excited, their energy must increase enough to cross the energy band gap and move to the bottom of the empty energy band higher in energy. After making this transition, the electron can increase its kinetic energy and be able to move in response to an electric field. Therefore, the energy band above the valence band is known as the conduction band and the electrons in the conduction band are known as conduction electrons, which serve as charge carriers in electrical conduction. In some insulators (e.g., diamond), the energy band gap is so large that it is unlikely for thermal energy to be sufficient to cause any of the valence electrons to move to the conduction band, so the concentration of conduction electrons is negligible in such an insulator. In some insulators, the low concentration of conduction electrons is due to the presence of carrier traps, such as defects. Because the carrier concentration is low, such a solid exhibits a low electrical conductivity. That is, it is an insulator.

2.3.3.3 Composites

A composite material is a material formed by the artificial blending of two or more components. For example, one component is metal particles while the other component is a polymer, such that the metal particles are dispersed in the polymer matrix. As polymers are usually insulators, a polymer-matrix composite that is electrically conductive usually contains an electrically conductive filler (particles, flakes, short fibers or continuous fibers).

In general, a composite material contains two or more components, which are usually different in electrical resistivity. The resistivity of the overall composite is related to the resistivity of each component, in addition to the proportions and distributions of the components.

Consider that the components are like wires parallel to one another, as illustrated in Fig. 2.9 for the case of a composite with three components (labeled 1, 2 and 3). The voltage gradient is in the direction of the "wires". Let the voltage difference between the two ends of the composite in the wire direction be V. Let the cross-sectional area of a component (i.e., all wires of the same component together) be A_1 for component 1, A_2 for component 2 and A_3 for component 3. Let

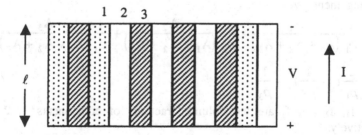

Fig. 2.9 Electrical conduction through a composite material consisting of three components (labeled 1, 2 and 3) that are in a parallel configuration.

the resistivities of the components be ρ_1, ρ_2 and ρ_3. The resistance R_i due to component i is thus given by

$$R_i = \rho_i \frac{\ell}{A_i}$$

Due to Ohm's law (Eq. (2.4)), the current I_i through component i is given by

$$I_i = \frac{VA_i}{\rho_i \ell}$$

Hence the total current through the composite is

$$I = I_1 + I_2 + I_3 = \frac{V}{\ell}\left(\frac{A_1}{\rho_1} + \frac{A_2}{\rho_2} + \frac{A_3}{\rho_3}\right) \qquad (2.21)$$

Now, let us consider the composite as a whole. Let the overall resistivity of the composite in the "wire" direction be $\rho_\parallel$ (the subscript signifying the parallel configuration). The total cross-sectional area is $A_1 + A_2 + A_3$. The total resistance is

$$R = \rho_\parallel \frac{\ell}{(A_1 + A_2 + A_3)}$$

Using Ohm's law, the total current I is given by

$$I = \frac{V}{R} = \frac{V(A_1 + A_2 + A_3)}{\rho_\parallel \ell} \qquad (2.22)$$

Combining Eq. (2.21) and (2.22), one obtains

$$\frac{V}{\ell}\left(\frac{A_1}{\rho_1} + \frac{A_2}{\rho_2} + \frac{A_3}{\rho_3}\right) = \frac{V}{\rho_\parallel \ell}(A_1 + A_2 + A_3)$$

Rearrangement gives

$$\frac{1}{\rho_\parallel} = \frac{1}{\rho_1}\frac{A_1}{(A_1 + A_2 + A_3)} + \frac{1}{\rho_2}\frac{A_2}{(A_1 + A_2 + A_3)} + \frac{1}{\rho_3}\frac{A_3}{(A_1 + A_2 + A_3)}$$

$$= \frac{1}{\rho_1}f_1 + \frac{1}{\rho_2}f_2 + \frac{1}{\rho_3}f_3, \qquad (2.23)$$

where f_1, f_2 and f_3 are the volume fractions of components 1, 2 and 3 respectively.

Next, consider that the components are in series, as illustrated in Fig. 2.10. The current I is the same through all the components. The cross-sectional area A perpendicular to the current is the same for all the components. However, the components differ in their lengths (L_1, L_2 and L_3) in the current direction. The resistance R_i in the current direction due to component i is given by

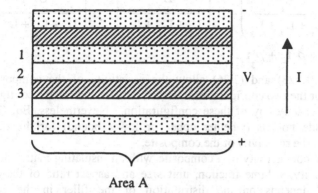

Area A

Fig. 2.10   Electrical conduction through a composite material consisting of three components (labeled 1, 2 and 3) that are in a series configuration.

$$V_i = IR_i$$

$$= I\rho_i \frac{L_i}{A}$$

The total voltage drop V is given by

$$V = \frac{I}{A}\left(\rho_1 L_1 + \rho_2 L_2 + \rho_3 L_3\right) \qquad (2.24)$$

Now let us consider the composite as a whole. Let the overall resistivity in the direction perpendicular to the long axis of each component be $\rho_\perp$. The total length is $L_1+L_2+L_3$. The total resistance R is given by

$$R = \rho_\perp \frac{\left(L_1 + L_2 + L_3\right)}{A}$$

Using Ohm's law, the total voltage drop is

$$V = IR$$

$$= I\rho_\perp \frac{\left(L_1 + L_2 + L_3\right)}{A} \qquad (2.25)$$

Combination of Eq. (2.24) and (2.25) and rearrangement give

$$\rho_\perp = \rho_1\left(\frac{L_1}{L_1+L_2+L_3}\right) + \rho_2\left(\frac{L_2}{L_1+L_2+L_3}\right) + \rho_3\left(\frac{L_3}{L_1+L_2+L_3}\right)$$

$$= \rho_1 f_1 + \rho_2 f_2 + \rho_3 f_3. \tag{2.26}$$

Eq. (2.23) and (2.26) allow the resistivity of the composite to be calculated for the two configurations. In the actual situation, the configuration is usually not exactly any of these configuration. Nevertheless, Eq. (2.23) and (2.26) provide bounds (whether upper or lower bound) for the purpose of estimation of the resistivity of the composite.

The conductivity of a composite with an insulating matrix depends on the conductivity, volume fraction, unit size and aspect ratio of the filler. In addition, it depends on the distribution of the filler in the composite. Thedistribution in turn depends on the composite fabrication method. In general, the conductivity of the composite does not increase linearly with the filler volume fraction. Rather, it increases abruptly at a certain filler volume fraction, which is called the percolation threshold. This threshold is the filler volume fraction at which the filler units begin to touch one another sufficiently so that somewhat continuous electrical conduction paths form in the composite (akin to liquid coffee drops touching one another and then liquid coffee flowing out of percolating coffee). Below and above the threshold, the conductivity still increases with the filler volume fraction, but more gradually. A large aspect ratio of the filler enhances the touching of adjacent filler units, thus decreasing the percolation threshold. In the case of a composite formed from a mixture of insulator matrix powder and filler units, the percolation threshold is lower when the ratio of the filler unit size to the matrix particle size is smaller, as the small filler units tend to line up along the boundaries of the large matrix particles.

The conductivity of a composite with an insulating matrix depends not only on the filler, but also on the filler-matrix interface. A stronger interface is associated with a lower contact resistivity, thus resulting in a lower volume resistivity for the composite. Below the percolation threshold, the filler-matrix interface plays a particularly important role. The contact resistivity (ρ_c) of an interface is a quantity that does not depend on the area of the interface or contact, but only depends on the nature of the interface. The contact resistance (R_c) decreases with increasing contact area (A), i.e.,

$$R_c \propto \frac{1}{A}$$

The proportionality constant is ρ_c, so that

$$R_c = \frac{\rho_c}{A} \tag{2.27}$$

Since the unit of R_c is Ω and that of A is m^2, the unit of ρ_c is $\Omega.m^2$.

2.3.3.4 Semiconductors

Semiconductors, like classical insulators, have completely filled valence bands. However, the energy band gap between the top of the filled valence band and the bottom of the conduction band is small, typically less than 4 eV. As a result, thermal energy causes a small fraction of the valence electrons to move across the energy band gap into the bottom of the conduction band, thus giving rise to a small number of conduction electrons, which serve as charge carriers for electrical conductions, as illustrated in Fig. 2.11.

The transition of a small fraction of the valence electrons from the top of the valence band to the bottom of the conduction band also gives rise to unoccupied electron states at the top of the valence band. This is analogous to the vacancies in a crystal. The movement of atoms can be described by the movement of vacancies in the opposite direction. Since the number of vacancies is a small fraction of the number of atom sites in the crystal, it is simpler to describe atomic relocation by the movement of the vacancies than by the movement of the atoms themselves. A similar situation applies to the electronic vacancies, called holes, in the top of the valence band of a semiconductor. Since the number of holes is much smaller than the number of electrons in the valence band, the transitions in energy of the valence electrons within the valence band [which is not completely filled, as shown in Fig. 2.11(b)] can more simply be described by the transitions in energy of the holes within the valence band. The holes provide an additional source of charge carriers. Since holes are due to the absence of electrons, they are positively charged. Hence, the charge of a hole is equal in magnitude and opposite in sign to that of an electron.

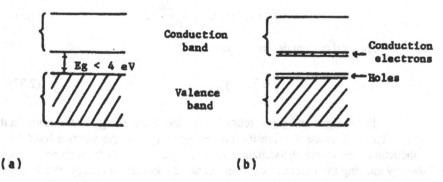

(a)   (b)

Fig. 2.11   Energy bands of a semiconductor. (a) At very low temperatures – negligible number of conduction electrons. (b) At higher temperatures – some conduction electrons. Shaded energy ranges are occupied by electrons.

Thus, a semiconductor has two types of charge carriers – conduction electrons and holes. The electrical conductivity of a semiconductor has two contributions, as given by

$$\sigma = q \, n \, \mu_n + q \, p \, \mu_p \ ,\tag{2.28}$$

where   q = magnitude of the charge of an electron,
          n = number of conduction electrons per unit volume,
          p = number of holes per unit volume,
          μ_n = mobility of conduction electrons,
and μ_p = mobility of conduction holes.

The subscript n is used for electrons, which are negatively charged; the subscript p is used for holes, which are positively charged. Since the transition of each electron from the valence band to the conduction band simultaneously gives rise to a conduction electron in the conduction band and a hole in the valence band, the number of conduction electrons must equal the number of holes, i.e., n = p. Therefore, for a semiconductor (pure), Eq. (2.28) can be simplified as

$$\sigma = qn(\mu_n + \mu_p) \ .\tag{2.29}$$

In the presence of both a voltage gradient and a charge carrier concentration gradient, the current density in a semiconductor has a contribution from the conduction electrons and another contribution from the holes. The contribution by the conduction electrons is given by

$$\tilde{J}_n = qn \, \mu_n \, E + qD_n \frac{dn}{dx} \ .\tag{2.30}$$

The contribution by the holes is given by

$$\tilde{J}_p = qp \, \mu_p \, E - qD_p \frac{dp}{dx} \ .\tag{2.31}$$

The total current density $\tilde{J}$ is given by

$$\tilde{J} = \tilde{J}_n + \tilde{J}_p \ .\tag{2.32}$$

The energy band gap is related to the ionization energy of the atoms in the solid. This is because the transition of an electron from the valence band to the conduction band means delocalizing and freeing the electron from its parent atom, thereby ionizing the parent atom. The higher the ionization energy, the larger the energy band gap. For example, consider the elements in group IV A of the periodic table. The ionization energies of these elements increse up the group, so the energy band gaps of these elements also increase up the group, as shown in Table 2.1. C (diamond), Si, Ge and Sn (gray) all exhibit the diamond structure. They all have completely filled valence bands which contain sp^3 hybridized electrons. The difference in electrical conductivity among these elements is due to the difference

Table 2.1   Energy band gaps and electrical conductivities of the elements in Group IVA of the periodic table.

Element	Energy band gap (eV)	Electrical conductivity at 20°C ($\Omega^{-1}.cm^{-1}$)
C (diamond)	~6	10^{-18}
Si	1.1	5×10^{-6}
Ge	0.72	0.02
Sn (gray)	0.08	10^4

in energy band gap. The gap is particularly large in diamond, so diamond is an insulator. The other three substances are all semiconductors. The smaller the energy band gap, the greater is the number of valence electrons that have sufficient thermal energy to cross the energy band gap, and the higher is the electrical conductivity.

Semiconductors have electrical conductivities that are between those of metals and those of insulators. Other than elemental semiconductors (e.g., Si, Ge, etc.), there are compound semiconductors (e.g., ZnS, GaP, GaAs, InP, etc.). In contrast to many elemental semiconductors, which exhibit the diamond structure, most compound semiconductors exhibit the zinc blend structure.

2.3.4   Temperature dependence of electrical resistivity

2.3.4.1   Metals

A metal has only one type of charge carriers – electrons. The electrical conductivity of a metal is given by Eq. (2.15):

$$\sigma = q\,n\,\mu \ ,$$

where   q = magnitude of the charge of an electron,
            n = number of free electrons (valence electrons) per unit volume,
and   μ = mobility of electrons.

The number of valence electrons does not change with temperature, so n is independent of temperature. On the other hand, as the temperature increases, the mobility of the electrons decreases. This is because thermal vibrations of the atoms in the solid increase in amplitude as the temperature increases and such vibrations interfere with the motion of the electrons, thus decreasing the mobility of the electrons. Since n is fixed and μ decreases slightly with increasing temperature, the electrical conductivity of a metal decreases slightly with increasing temperature. In other words, the electrical resistivity of a metal increases with increasing temperature. The variation of the electrical resistivity with temperature is roughly linear, as shown schematically in Fig. 2.12. The rate of increase of electrical resistivity with temperature is commonly described by the temperature coefficient of electrical resistivity (α), which is defined by the equation

$$\frac{\Delta\rho}{\rho} = \alpha\Delta T , \tag{2.33}$$

where $\Delta\rho/\rho$ is the fractional change in electrical resistivity, and ΔT is the increase in temperature. The value of α is ~0.004°C^{-1} for most pure metals. Note that Eq. (2.33) requires that the unit of α be °C^{-1}, which is the same as K^{-1}, where K means Kelvin or absolute temperature.

2.3.4.2 Semiconductors

A semiconductor has two types of carriers -- conduction electrons and holes. The electrons (negative) move up a voltage gradient, whereas the holes (positive) move down a voltage gradient, as illustrated in Fig. 2.1. Conventionally, the direction of the current is taken as the direction of the flow of the positive charge carriers, i.e., from a higher voltage to a lower voltage.

The electrical conductivity of an intrinsic semiconductor is given by Eq. (2.29), which is

$$\sigma = q \, n \, (\mu_n + \mu_p) \; . \tag{2.34}$$

Due to thermal vibrations of the atoms, both μ_n and μ_p decrease slightly as the temperature increases. Since thermal energy causes the valence electrons to be excited to the conduction band, the value of n increases with increasing

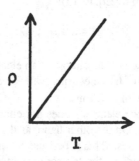

Fig. 2.12   The temperature dependence of the electrical resistivity of a metal.

temperature. As for any thermally activated process, the temperature dependence is exponential. It can be shown that n varies with temperature in the form

$$n \propto e^{-E_g/2kT},$$

where E_g = energy band gap between conduction and valence bands,

k = Boltzmann's constant,

and T = temperature in K.

The factor of 2 in the exponent is because the excitation of an electron across E_g produces an intrinsic conduction electron and an intrinsic hole.

Since μ_n and μ_p only decrease slightly with increasing temperature, whereas n significantly increases with increasing temperature, the variation of σ with temperature is roughly of the form,

$$\sigma \propto e^{-E_g/2kT}.$$

E_g is itself temperature-dependent, although its slight temperature dependence is negligible in most circumstances. Let the constant of proportionality be σ_0.

Taking natural logarithms,

$$\sigma = \sigma_0 e^{-E_g/2kT}.$$

$$\ln\sigma = \ln\sigma_0 - \frac{E_g}{2kT}. \tag{2.35}$$

Changing the natural logarithms to logarithms of base 10,

$$\log\sigma = \log\sigma_0 - \frac{E_g}{(2.3)2kT}. \tag{2.36}$$

Thus, the variation of σ with T can be shown in an Arrhenius plot of log σ against 1/T, as shown in Fig. 2.13. The plot is a straight line of slope $-[E_g/(2.3).2k]$; the intercept of the line on the log σ axis is log σ_0. Therefore, by measuring σ as a

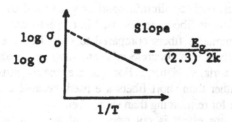

Fig. 2.13 An Arrhenius plot of the variation of the electrical conductivity with temperature.

function of T, one can determine E_g and σ_o. This is one of the most common ways to determine the energy band gap of a semiconductor.

The significant variation of σ with T for a semiconductor allows a semiconductor to serve as a temperature measuring device called a thermistor. The temperature is indicated by the conductivity, which relates to the measured resistance of the thermistor. The resistance decreases with increasing temperature.

The most clearcut difference in behavior between a metal and a semiconductor is that the electrical conductivity of a metal decreases with increasing temperature whereas that of a semiconductor increases with increasing temperature. Therefore, the simplest way to determine whether a solid is a metal or a semiconductor is to see whether the electrical conductivity of the solid increases or decreases with temperature.

2.3.4.3 Composites

A new class of thermistors involves carbon fiber structural composites instead of semiconductors [1]. The advantages of the composites are low cost, mechanical rugged and processability into various shapes and sizes (as small as a coin or as large as a bridge). The processability into large sizes means that a composite structure is inherently able to sense its own temperature, without the need for embedded or attached sensors. This makes the structure intrinsically smart. Compared to the use of embedded or attached sensors, the intrinsically smart structure has the advantages of low cost, high durability, large sensing volume (everywhere rather than just here and there) and absence of mechanical property degradation (which occurs in the case of embedded sensors).

Concrete is a cement-matrix composite that is important for civil structures. The addition of short fibers as an admixture is known to decrease the drying shrinkage and increase the flexural toughness. In the case of the fibers being carbon fibers, the fiber addition also increases the flexural strength and renders the composite the ability to sense its own strain (Sec. 2.5). The strain sensing ability is due to the effect of strain on the volume electrical resistivity of the composite.

Continuous carbon fiber polymer-matrix composites are widely used for lightweight structures, such as aircraft, sporting goods and even automobiles and wheel chairs. Continuous fibers are not used in cement-matrix composites due to the high cost of continuous fibers compared to short fibers, the impossibility of having continuous fibers in a concrete mix, and the importance of low cost for a concrete to be industrially viable. For polymer-matrix structural composites, continuous fibers rather than short fibers are used, because continuous fibers are much more effective for reinforcing than short fibers.

The thermistor effect is not only useful for application in temperature sensing, it is relevant to the study of the electrical conduction mechanism, which relates to the structure of the composite. The sensitivity of a thermistor for temperature sensing is described by the activation energy of the electrical conduction, as obtained from the negative slope of the Arrhenius plot of the logarithm of the electrical conductivity versus the reciprocal of the absolute temperature. The activation energy reflects the energy for the hopping of the charge carrier in the composite.

This section describes the thermistor effect in short carbon fiber cement-matrix composite and continuous carbon fiber polymer-matrix composite. The former is associated with an activation energy that is similar to the values for semiconductors, whereas the latter is associated with a lower activation energy. Hence, the former is superior to the latter for thermistor application. However, the latter is more amenable to practical implementation, because the continuous fibers serve as electrical leads, so that both thermistors and leads are built into a continuous fiber composite.

Carbon fiber silica fume cement paste is an effective thermistor. The electrical resistivity decreases with increasing temperature (1-45°C), with an activation energy of electrical conduction (electron hopping) of 0.4 eV, which is comparable to those of semiconductors (typical thermistor materials) and higher than that of carbon fiber polymer-matrix composites. Without carbon fibers, or with latex in place of silica fume, the activation energy is much lower and the resistivity is higher. The voltage range for linear current-voltage characteristic is narrower when fibers are present than when fibers are absent. Linearity occurs up to 8 V for carbon fiber silica fume cement paste at 20°C.

An epoxy-matrix continuous carbon fiber composite comprising two crossply laminae is a thermistor array. Each junction between crossply fiber tow groups of the adjacent laminae is a thermistor, while the fiber groups serve as electrical leads (Fig. 2.14). The contact electrical resistivity of the junctions decreases reversibly upon heating, due to the electron hopping between the

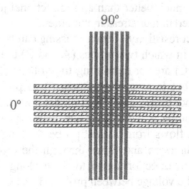

Fig. 2.14   Thermistor array in the form of a carbon fiber polymer-matrix composite comprising two crossply laminae.

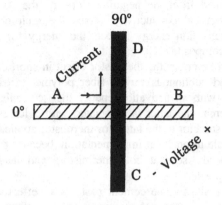

Fig. 2.15   A single thermistor in the form of a junction between two crossply laminae. The contact resistivity of the junction is measured by using current probes A and D and voltage probes B and C. The fibers serve as electrical leads.

laminae. The fractional change in contact resistivity provides an indication of temperature. The contact resistivity decreases with increasing pressure during composite fabrication, due to increase in pressure exerted by fibers of one lamina on those of the other lamina. The magnitude of the fractional change in contact resistivity per degree C increases with increasing curing pressure (fiber volume fraction), due to the increase in interlaminar stress with increasing fiber volume fraction and the consequent increase in activation energy (up to 0.12 eV). A crossply junction is much better than a unidirectional junction as a thermistor, due to the absence of interlaminar stress in the latter.

The contact resistivity of each crossing can be best measured by using the four-probe method, in which two probes (A and D) are for passing the current and two probes (B and C) are for measuring the voltage (Fig. 2.15). (The two-probe method is simpler but less accurate; it uses two probes so that the current and voltage probes are not separate and the contact potential resulting from the contact resistance of each probe is included in the measured voltage.) In the four-probe method, the current flows from current probe A along one fiber tow, turns to the through-thickness direction and flows through the crossing from one tow to the other, and then turns direction again to flow along the other fiber tow toward current probe D. The voltage between probes B and C gives the voltage across the junction. The resistance multiplied by the junction area is the contact resistivity of the junction. The array of junctions in Fig. 2.14 allows determination of the temperature distribution. Computer data acquisition allows the contact resistivity of the junctions in the array to be measured one junction at a time, so that the whole array is measured in a reasonably short time. As the fiber tows serve as electrical leads, no wiring is needed. Not all the tows need to serve as leads, because the spatial resolution of the sensing does not need to be excessive. Moreover, a group of tows (rather than a single tow) can serve as a lead.

The top two fiber layers of a composite structure capable of temperature sensing should be crossply (Fig. 2.14). The layers below can be in other lay-up configurations. The fibers in the top two layers should be longer than those in the other layers in order to facilitate electrical connection.

A thermocouple array can be used to provide spatially resolved light detection. However, a thermocouple array requires much wiring. In addition, the thermocouple tips must be at or near the outer surface of the composite structure. If the thermocouples are embedded in the composite, they are intrusive and degrade the mechanical properties of the composite. If the thermocouples are attached on the surface of the composite, they can come off easily. Therefore, in practice, a thermocouple array is not feasible for spatially resolved light detection.

2.3.5 Ionic conduction

Cement-based materials are partly ionic. In an ionic solid, the energy band gap is usually very large, so that there are very few conduction electrons. The predominant charge carriers in an ionic solid are frequently the ions, which move by diffusion. The diffusion process most often proceeds by the vacancy mechanism, which requires the presence of vacancies. For electrical neutrality, the vacancies in an ionic solid typically occur in pairs of cation and anion vacancies. In other words, rather than single vacancies, an ionic solid more frequently has Schottky defects. The movement of the ions can be described by that of the Schottky defects, just as the movement of atoms can be described by that of the vacancies. The number of charge carriers is thus determined by the number of Schottky defects, rather than the number of ions in the solid.

In general, an ionic solid has two types of charge carriers - cations and anions. The cations are positively charged, so they move down a voltage gradient; the anions are negatively charged, so they move up a voltage gradient. Since anions are usually larger than cations, the diffusion coefficient of cations is usually higher than that of anions. Therefore, cations are usually the dominant charge carriers in an ionic solid.

The electrical conductivity of an ionic solid is given by

$$\sigma = q\,n\,\mu_C + q\,n\,\mu_A = q\,n\,(\mu_C + \mu_A) \ , \qquad (2.37)$$

where $\quad$ n = number of Schottky defects per unit volume

μ_C = mobility of cations,

μ_A = mobility of anions.

The mobilities μ_c and μ_A are determined by the diffusion coefficients of the cations and anions, respectively, in the solid. The diffusion coefficient of an ion increases with increasing temperature, such that

$$D \propto e^{-E/kT},$$

where E = activation energy for the relocation of an ion. Neglecting the factor of T which relates the mobility to the diffusion coefficient [Eq. (2.20)], the mobilities roughly increase exponentially with temperature in this form.

As for vacancy formation, the formation of Schottky defects involves a certain activation energy ΔH. Therefore, n increases with temperature in the form

$$n \propto e^{-\Delta H/kT}, \tag{2.38}$$

where ΔH = activation energy for the formation of a Schottky defect.

Since both the mobilities and n increase roughly exponentially with increasing temperature, the electrical conductivity, as given by Eq. (2.39), also increases roughly exponentially with temperature:

$$\sigma \propto e^{-(\Delta H+E)/kT}. \tag{2.39}$$

Hence, the temperature dependence of the electrical conductivity of an ionic solid can also be shown by means of an Arrhenius plot of log σ against 1/T.

2.3.6   Semiconductors

2.3.6.1   Intrinsic semiconductors

Intrinsic semiconductors are pristine or pure semiconductors.   The electrical conductivity of intrinsic semiconductors is an inherent property of these materials and does not arise from impurities.   Si and Ge are the two most well-known intrinsic semiconductors.   In terms of the energy band model, we have described in Section 2.3.3.4 how charge carriers are generated in an intrinsic semiconductor.   An alternative way of telling the same story is by means of the valence bond model, which is described below.

2.3.6.1.1   Generation of carriers

Consider, for example, Si, which exhibits the diamond structure.   In Si, the atoms are sp^3 hybridized, so that every atom is tetrahedrally coordinated with four other atoms by covalent bonding.   A projected two-dimensional view of the crystal structure is shown in Fig. 2.16(a).   Each Si atom has four valence electrons, as indicated by the dot notation.   The transition of a valence electron from the top of the valence band to the bottom of the conductin band is the same as having a valence electron gain an amount of energy equal to E_g and thereby break loose from an Si-Si covalent bond, so that the electron becomes free to move about and serve as a conduction electron, as illustrated in Fig. 2.16(b).   The energy band gap represents the amount of energy required to free a valence electron from its parent

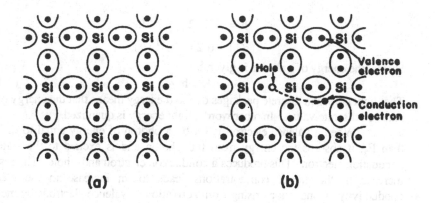

Fig. 2.16   Intrinsic semiconduction in Si.

atom. The freeing of the electron produces a hole at the original position of the valence electron, as indicated by an open circle in Fig. 2.16(b). In the energy band model, we say that the transition of an electron from the valence band to the conduction band produces a hole in the valence band, in addition to an electron in the conduction band. The hole produced can move to the position of an electron in another Si-Si bond by having the electron there move to take the position of the hole. Thus, the movement of a hole is analogous to that of a vacancy. In this fashion, the hole serves as a positive charge carrier.

Thermal energy is the most common cause of exciting a valence electron to become a conduction electron. The process by which thermal agitation creates a conduction electron and hole is called thermal ionization. The higher the temperature, the higher the fraction of valence electrons that are excited.

Another method of carrier generation is to shine light on the semiconductor of energy equal to or higher than that of the energy band gap (E_g).

Light is electromagnetic radiation which can be considered as a wave of wavelength λ. The wavelength of visible light ranges from 7×10^{-5} cm (red light) to 4×10^{-5} cm (violet light). The speed of light (c) is 3×10^{10} cm.s^{-1}. The frequency (v) of the wave is related to c and λ by

$$v = \frac{c}{\lambda} .$$ (2.40)

The unit of v is s^{-1} (i.e., cycles per second), which is also called Hertz (Hz).

Light can also be considered as quanta (packages) called photons, such that the energy of a photon E is proportional to v.

$$E \propto v .$$

The constant of proportionality is called the Planck's constant (h), where

$$h = 6.6262 \times 10^{-34} \text{ J.s.}$$

$$(1 \text{ J} = 6.24 \times 10^{18} \text{ eV}).$$

Hence, the energy of a photon is given by

$$E = h\nu \ . \tag{2.41}$$

That light comes in discrete packages of fixed energy means that the energy of light takes on discrete values. In other words, light energy is quantized.

When a valence electron is hit by a photon of energy equal to or higher than E_g, it absorbs the energy from the photon and is excited to the state of a conduction electron. This produces a conduction electron and a hole. The resulting increase in the carrier concentrations leads to an increase in the electrical conductivity. Conduction arising from activation of valence electrons by means of light is known as photoconduction.

2.3.6.1.2   Recombination of carriers

Since energy is needed to produce each pair of conduction electron and hole, energy is released if the electron-hole pair recombines. In other words, in recombination, a hole and a conduction electron meet and annihilate each other by placing the electron in the position of the hole (which is a vacant valence electron site). Thus, the conduction electron becomes a valence electron and the hole disappears.

The recombination rate is proportional to the concentration of carriers available for recombination, and it can be expressed as

$$\text{Recombination rate} = R = apn \ , \tag{2.42}$$

where a is a constant accounting for the material properties of the semiconductor. Eq. (2.42) applies to band-to-band recombination, where an electron in the conduction band and a hole in the valence band recombine directly without the aid of intermediate centers (energy levels in the band gap due to imperfections in the semiconductor).

For an intrinsic semiconductor, $n = p$. Let $n = p = n_i$, where n_i denotes the concentration of holes or conduction electrons in an intrinsic semiconductor. Hence, for an intrinsic semiconductor, Eq. (2.42) becomes

$$R = an_i^2 \tag{2.43}$$

Thermal ionization causes the concentration of holes and conduction electrons to increase until, at thermal equilibrium, the recombination rate R equals the ionization rate I. Thus, at equilibrium,

$$I = R = an_i^2$$

Note that $I = I(T)$ is a strong function of the temperature T.

2.3.6.2   Extrinsic semiconductors

An intrinsic semiconductor contains equal concentrations of conduction electrons and holes (i.e., n = p). Extrinsic semiconductors, also known as doped semiconductors, contain impurities or defects which introduce excess conduction electrons or excess holes, so that the concentrations of conduction electrons and holes are not equal (i.e., n ≠ p). The excess carriers serve to increase the electrical conductivity of the semiconductor. If n > p, the extrinsic semiconductor is said to be n-type; if n < p, the extrinsic semiconductor is said to be p-type. Depending on the impurity or defect in the semiconductor, the excess carriers can be electrons or holes, so that the extrinsic semiconductor can be n-type or p-type.

2.3.6.2.1   Impurity semiconductors

In all extrinsic semiconductors, the impurity atoms reside at the atom sites of the semiconductor crystal. In other words, the impurity is a substitutional impurity.

2.3.6.2.1.1   n-type semiconductors

Consider, for example, silicon (Si) containing a small amount of phosphorous (P) as a substitutional impurity. That is, the Si-P alloy is a substitutional solid solution with a small amount of P in Si. The alloy exhibits the diamond structure of Si. A projected two-dimensional view of the crystal structure is shown in Fig. 2.17(a). Phosphorous is in group V A of the periodic table, whereas silicon is in group IV A. Thus, P has five valence electrons and Si has

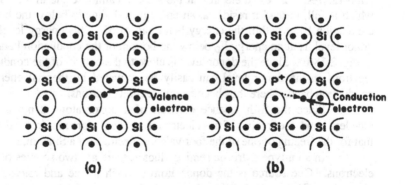

(a)   (b)

Fig. 2.17   N-type extrinsic semiconductor due to an impurity.

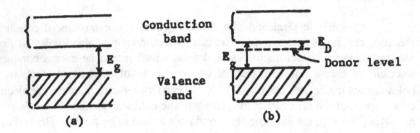

Fig. 2.18   The energy bands of (a) an intrinsic semiconductor and (b) an n-type extrinsic semiconductor.

four. Four of the five valence electrons of a P atom are engaged in forming four σ covalent bonds with four neighboring Si atoms, as shown in Fig. 2.17(a) by means of the dot notation.   By absorbing a certain amount of energy, the remaining valence electron of P is freed from the P atom, resulting in a conduction electron and a P^+ cation, as illustrated in Fig. 2.17(b). In this fashion, P provides excess conduction electrons, so that the concentration of conduction electrons becomes higher than that in pristine Si. As a result, n > p. Therefore, P is said to be an electron donor in Si; it makes the semiconductor n-type.

The above description of the behavior of an electron donor is in terms of the valence bond model. The same story can be told in terms of the energy band level. The conduction and valence bands of Si are shown in Fig. 2.18(a). The valence band is full at very low temperatures and is almost full at higher temperatures. The extra electron at the P atom cannot reside in the valence band, which is full; instead it resides at an energy level slightly below the bottom of the conduction band within the energy band gap of Si. This energy level, called the donor level, is at an energy E_D below the bottom of the conduction band, as shown in Fig. 2.18(b). Since the donor level is close to the bottom of the conduction band, electrons in the donor level can easily be activated by thermal energy into the conduction band, where they become conduction electrons.

Since the fifth valence electron of a donor atom is not engaged in a covalent bond, the freeing of this electron from the P atom requires less energy than that for the freeing of one of the four valence electrons of a Si atom, i.e., $E_D < E_g$.

In an n-type extrinsic semiconductor, there are two sources of conduction electrons. One source is the donor atoms, which ionize and release conduction electrons.   These conduction electrons which arise from the impurity in the semiconductor are known as extrinsic conduction electrons. The other source of conduction electrons are the host atoms, which release valence electrons and thereby produce equal numbers of conduction electrons and holes.   These conduction electrons and holes which arise from the intrinsic semiconductor are known as intrinsic conduction electrons and intrinsic holes. Thus, the concentration of conduction electrons in an n-type extrinsic semiconductor has two contributions:

$$n = n_i + n_e , \tag{2.44}$$

where n = total concentration of conduction electrons,

n_i = concentration of intrinsic conduction electrons,

n_e = concentration of extrinsic conduction electrons.

The generation of an extrinsic conduction electron is accompanied by the change of a donor atom into a cation. This process can be represented by the equation

$$D \rightarrow D^+ + e^- ,$$

where D denotes a donor atom and e^- denotes an electron. Therefore,

$$n_e = N_D + , \tag{2.45}$$

where N_D+ is the concentration of donor ions. Because the number of donor atoms in solid solution is limited, the maximum value of n_e is limited to N_D (the concentration of donor atoms in solid solution). When all the donor atoms have ionized, n_e has reached its maximum, and this situation is known as donor exhaustion.

Since thermal energy is the most common energy source for the excitation of valence electrons to become conduction electrons, the relative importance of n_i and n_e depends on the temperature. For the case of an intrinsic semiconductor,

$$n_i \propto e^{-Eg/2kT} . \tag{2.46}$$

On the other hand, extrinsic conduction electrons are produced by excitation across an energy of E_D. Therefore,

$$n_e \propto e^{-E_D/kT} . \tag{2.47}$$

Since $E_g/2 > E_D$, comparison of Eqs. (2.46) and (2.47) indicates that, when the donor atoms are not exhausted,

$$n_i << n_e . \tag{2.48}$$

However, at high temperatures where all the donor atoms have ionized, n_e is limited and Eq. (2.48) is not necessarily valid.

In an n-type extrinsic semiconductor, there are no extrinsic holes, so

$$p = p_i . \tag{2.49}$$

Since $p_i = n_i$, Eq. (2.49) becomes

$$p = n_i .$$

Thus, when Eq. (2.48) is valid, Eqs. (2.44) can be written, respectively, as

$$n \cong n_e \tag{2.50}$$

and

$$p \cong 0 . \tag{2.51}$$

The electrical conductivity of a semiconductor is given by Eq. (2.28):

$$\sigma = qn\,\mu_n + qp\,\mu_p \ . \tag{2.28}$$

For an n-type extrinsic semiconductor, when Eqs. (2.50) and (2.51) are valid (i.e., before donor exhaustion),

$$\sigma \cong qn\,\mu_n \ ,$$

where the variation of n with temperature is given by Eq. (2.47). Thus, an Arrhenius plot of log σ versus 1/T gives a straight line of slope $-[E_D/2.3\,k]$. This temperature dependence of the electrical conductivity is only valid before donor exhaustion, as shown in the low temperature range of Fig. 2.19. As the temperature increases, more and more of the donor atoms are ionized. When donor exhaustion occurs, n_e no longer increases with temperature and the electrical conductivity does not vary much with temperature, as indicated in the intermediate temperature range of Fig. 2.19. Although n_e is fixed, n_i keeps on increasing with temperature, according to Eq. (2.46), since there are many more host atoms than donor atoms in the semiconductor. Therefore, at high enough temperatures, n_i can exceed n_e to the point where $n_i \gg n_e$. Under this situation, Eq. (2.44) becomes

$$n \cong n_i \tag{2.52}$$

where n_i varies with temperature in the form of Eq. (2.40). Thus, an Arrhenius plot at these high temperatures is a straight line of slope $-[E_g(2.3)2k]$, as for an intrinsic semiconductor. This plot is shown in the high temperature range of Fig. 2.19. Elements from group V A (N, P, As and Sb) of the periodic table can all act as electron donors to semiconductors in group IV A (Si, Ge).

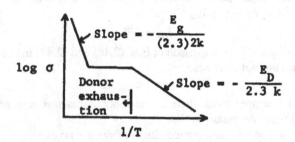

Fig. 2.19   Temperature dependence of electrical conductivity for an n-type semiconductor. Note the difference in slope between the two slanted portions of the plot, because $E_g \gg E_D$.

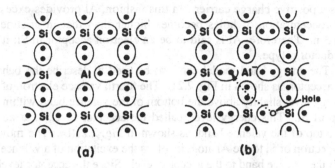

Fig. 2.20 P-type extrinsic semiconductor due to an impurity.

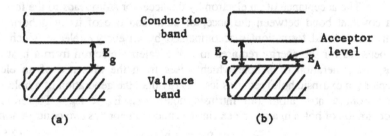

Fig. 2.21 The energy bands of (a) an intrinsic semiconductor and (b) a p-type extrinsic semiconductor.

2.3.6.2.1.2 p-type semiconductors

Consider Si containing a small amount of aluminum (Al) as a substitutional impurity. A projected two-dimensional view of the crystal structure is shown in Fig. 2.20(a). Aluminum is in group III A of the periodic table, whereas silicon is in group IV A. Thus, Al has three valence electrons and Si has four. Therefore, an Al atom can form only three σ bonds with three neighboring Si atoms, as shown in Fig. 2.20(a) by means of the dot notation. Thermal energy may cause a valence electron of a nearby Si atom to move to the Al atom, so that the Al atom becomes an Al⁻ anion and the Al⁻ anion has a fourth valence electron to form a σ bond with the remaining neighboring Si atom, as shown in Fig. 2.20(b). This movement causes a hole at the Si atom from which the electron moves. The hole can move from one Si atom to another, and thus serves as a positive charge carrier. In this fashion, Al provides excess holes, so that the concentration of holes becomes higher than that in pristine Si. As a result, $p > n$. Therefore, Al is said to be an electron acceptor in Si; it makes the semiconductor p-type.

The energy band model can also be used to describe the behavior of an electron acceptor, as shown in Fig. 2.21. The fourth valence electron of Al^- resides at an energy level slightly above the bottom of the valence band within the energy band gap of Si. This energy level, called the acceptor level, is at an energy E_A above the top of the valence band, as shown in Fig. 2.21(b). The movement of a valence electron of Si to the Al atom involves the excitation of a valence electron at the top of the valence band to the acceptor level. Since the acceptor level is close to the top of the valence band, this excitation requires much less energy than that needed to excite an electron from the valence band across the energy band gap to the conduction band. When the electron in the valence band has been excited to the acceptor level, a hole is produced in the valence band.

The acceptance of an electron by the acceptor atom leads to the formation of a covalent bond between the acceptor atom and one of its neighboring host atoms. The bond formation is accompanied by an energy release which partly compensates for the energy required to free a valence electron from a host atom. Thus, the generation of a hole which arises from the presence of an electron acceptor (an extrinsic hole) requires less energy than the generation of a hole which arises from a host atom (an intrinsic hole); i.e., $E_A < E_g$. Therefore, the concentration of holes in a p-type extrinsic semiconductor has two contributions:

$$p = p_i + p_e \ , \tag{2.53}$$

where p = total concentration of conduction holes
                p_i = concentration of intrinsic holes,
                p_e = concentration of extrinsic holes.

The generation of an extrinsic hole is accompanied by the change of an acceptor atom into an anion. This process can be represented by the equation

$$A + e^- \rightarrow A^- \ ,$$

where A denotes an acceptor atom. Alternatively, this equation can be written as

$$A \rightarrow A^- + h^+ \ ,$$

where h^+ denotes a hole. Therefore,

$$p_e = N_{A^-} \ , \tag{2.54}$$

where N_{A^-} is the concentration of acceptor ions. Because the number of acceptor atoms in solid solution is limited, the maximum value of p_e is limited to N_A, the concentration of acceptor atoms in solid solution. When all the acceptor atoms have accepted electrons, p_e has reached its maximum, and this situation is known as acceptor saturation.

The relative importance of p_i and p_e depends on the temperature. As for n_i,

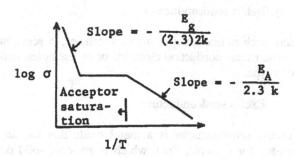

Fig. 2.22   Temperature dependence of electrical conductivity for a p-type semiconductor. Note the difference in slope between the two slanted portions of the plot, because $E_g \gg E_A$.

$$p_i \propto e^{-E_g/2kT} , \qquad (2.55)$$

and, as in Eq. (2.47) for n_e,

$$p_e \propto e^{-E_A/kT} . \qquad (2.56)$$

Since $E_g/2 > E_A$, a comparison of Eq. (2.55) and (2.56) indicates that, before the acceptor atoms are saturated,

$$p_i \ll p_e . \qquad (2.57)$$

However, at high temperatures where all the acceptor atoms have received electrons from the valence band, p_e is limited and Eq. (2.57) is not necessarily valid. A p-type extrinsic semiconductor has no extrinsic conduction electrons, so

$$n = n_i . \qquad (2.58)$$

Since $n_i = p_i$, Eq. (2.58) becomes

$$n = p_i .$$

When Eq. (2.57) is valid, Eq. (2.53) and (2.58) can be written, respectively, as

$$p \cong p_e \qquad (2.59)$$

and

$$n \cong 0 . \qquad (2.60)$$

A p-type semiconductor exhibits a temperature dependence of electrical conductivity similar to that of an n-type semiconductor, as shown in Fig. 2.22.

Elements from group III A (B, Al, Ga and In) of the periodic table can all act as electron acceptors to semiconductors in group IV A (Si, Ge).

2.3.6.2.2   Defect semiconductors

Defects, such as interstitial atoms or vacancies, in semiconducting metal oxides can provide excess conduction electrons or excess holes, making the metal oxide n-type or p-type.

2.3.6.2.2.1   Excess semiconductors

An excess semiconductor is a metal oxide that has an excess metal content. Consider, for example, ZnO, which is an ionic solid containing Zn^{2+} and O^{2-} ions. When exposed to a reducing atmosphere so that a small fraction of the O^{2-} ions are removed, ZnO becomes $Zn_{1+x}O$, where $0 < x \ll 1$. A stoichiometric compound is one that contains atoms of different elements in integral proportions; otherwise, the compound is said to be nonstoichiometric. ZnO is stoichiometric, whereas $Zn_{1+x}O$ is nonstoichiometric. Because Zn^{2+} ions can move more easily than O^{2-} ions, $Zn_{1+x}O$ contains zinc interstitial ions rather than O^{2-} vacancies, as shown in Fig. 2.23. The presence of an interstitial zinc ion can be considered as having two zinc ions at a Zn^{2+} site. For electrical neutrality, the two zinc ions at a Zn^{2+} site must together have a charge of +2. This requires that each of them be of charge +1, so that the two ions are each a Zn^+ ion. A Zn^+ ion has one electron more than a Zn^{+2} ion, so it can serve as an electron donor. By donating an electron, the Zn^+ ion becomes a Zn^{2+} ion: $Zn^+ \rightarrow Zn^{2+} + e^-$.

The situation can also be described by using the energy band model (Fig. 2.24). The extra electron on each Zn^+ ion resides at the donor level, which is close to the bottom of the conduction band. When thermal energy causes a Zn^+ ion to change into a Zn^{2+} ion, an electron is excited from the donor level to the bottom of the conduction band, where it becomes a conduction electron. Therefore, $Zn_{1+x}O$ is n-type.

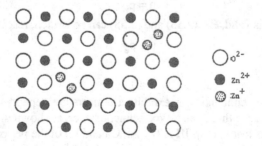

Fig. 2.23   A section of the crystal structure of $Zn_{1+x}O$.

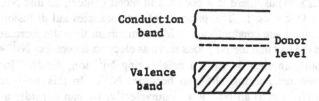

Fig. 2.24 Energy bands of $Zn_{1+x}O$.

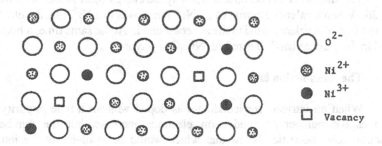

Fig. 2.25 A section of the crystal structure of $Ni_{1-x}O$.

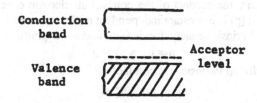

Fig. 2.26 Energy bands of $Ni_{1-x}O$.

2.3.6.2.2.2 Deficit semiconductors

A deficit semiconductor is a metal oxide that has a decreased metal content.

Consider, for example, NiO, which is an ionic solid containing Ni^{2+} and O^{2-} ions. When NiO is oxidized, some of the Ni^{2+} ions become Ni^{3+} ions. However, for electrical neutrality, three Ni^{2+} ions are replaced by two Ni^{3+} ions and a vacancy, as illustrated in Fig. 2.25. Thus, there is a decrease in metal content, so that NiO becomes $Ni_{1-x}O$, where $0 < x \ll 1$. The presence of the vacancies aid diffusion, and therefore increases the ionic conductivity. More important than the increase in ionic conductivity is the fact that the Ni^{3+} ions serve as electron acceptors: $Ni^{3+} + e^- \rightarrow Ni^{2+}$. By accepting an electron from a neighboring Ni^{2+} ion, the Ni^{3+} ion becomes Ni^{2+}, while the neighboring Ni^{2+} ion becomes Ni^{3+}. In this way, an electron hops from an Ni^{2+} ion to an Ni^{3+} ion. Equivalently, we can consider an Ni^{3+} ion to release a hole (h^+) when it becomes Ni^{2+}, i.e., $Ni^{3+} \rightarrow Ni^{2+} + h^+$. The hole moves from an Ni^{3+} ion to an Ni^{2+} ion, turning the Ni^{3+} to Ni^{2+} and the Ni^{2+} to Ni^{3+}.

The situation can also be described by the energy band model, as shown in Fig. 2.26. When a valence electron of an Ni^{2+} ion moves to a Ni^{3+} ion, the electron is excited from the valence band to the acceptor level. At the same time, a hole is created in the valence band. Therefore, $Ni_{1-x}O$ is p-type.

2.3.6.3 The mass-action law

When an intrinsic semiconductor is doped with an n-type impurity, not only does the number of conduction electrons increase, but the number of conduction holes decreases below that which would be available in the intrinsic semiconductor. The reason for the decrease in the number of holes is that the larger number of electrons present increases the rate of recombination of electrons with holes. Similarly, doping with p-type impurities decreases the concentration of free electrons below that in the intrinsic semiconductor. It can be shown that, under thermal equilibrium, the product of the conduction electron concentration (n) and hole concentration (p) is a constant independent of the amount of donor or acceptor impurity. For an intrinsic semiconductor (i.e., without doping),

$$n = n_i = p_i = p .$$

This means that, for an intrinsic semiconductor,

$$np = n_i^2 .$$

Since np is a constant independent of the doping level, the constant is just n_i^2, where n_i is the intrinsic carrier concentration. Thus, for both intrinsic and extrinsic semiconductors,

$$np = n_i^2 . \qquad (2.61)$$

Note that n_i depends only on the temperature and the material. At room temperature (300 K),

$$n_i = 1.5 \times 10^{10} \, cm^{-3} \; for \, Si$$

and

$$n_i = 2.5 \times 10^{13} \, cm^{-3} \, for \, Ge .$$

Eq. (2.61) is known as the mass-action law. It is used for obtaining the concentration of the holes when that of the conduction electrons is known and for obtaining the concentration of the conduction electrons when that of the holes is known. For example, consider an n-type semiconductor with a donor atom concentration of N_D. At ordinary temperatures,

$$n \cong n_e = N_{D+} \tag{2.62}$$

Suppose that the temperature is not high enough to cause n_e to be comparable to n_i, but is sufficient to cause donor exhaustion. In fact, this condition is often satisfied at room temperature. Then

$$n_{D+} = n_D$$

so that Eq. (2.50) becomes

$$n \cong N_D .$$

By using the mass-action law, the hole concentration is then given by

$$p = \frac{n_i^2}{n} = \frac{n_i^2}{N_D} .$$

2.3.6.4 Majority and minority carriers

An extrinsic semiconductor usually has a much higher concentration of one type of carrier than the other. The carriers of higher concentration are called the majority carriers; those of lower concentration are called the minority carriers. Thus, in an n-type semiconductor, the conducting electrons are the majority carriers, whereas the holes are the minority carriers.

The low concentration of minority carriers makes this quantity very easily disturbed by recombination. On the other hand, majority carriers are so abundant that slight changes its concentration due to recombination or carrier injection do not have a significant effect on the majority carrier concentration. As a result, the minority carriers often control the behavior of a semiconductor device.

Fig. 2.27 shows a thin slab of an n-type semiconductor that is illuminated with light capable of ionizing the atoms in the semiconductor and generating electron-hole pairs. It is assumed that the slab is so thin that the light passes completely through it and illuminates the whole volume of the slab uniformly. Thus, photo-ionization occurs uniformly throughout the slab, and the concentration of minority-carrier holes is increased uniformly above the thermal equilibrium level. Under these conditions the concentration of holes in the slab can be written as

$$p_n = p_{no} + p_n' . \tag{2.63}$$

where p_{no} is the thermal equilibrium concentration and p_n' is the excess concentration caused by the light. The subscript n is used to emphasize the fact that the semiconductor is n-type. According to Eq. (2.42), the recombination rate is $R = a p_n n_n$, and the carrier concentrations build up to the level at which the recombination rate equals the total ionization ratedue to the thermal and photo

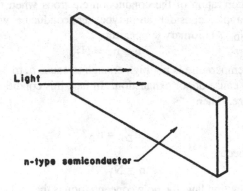

Light

n-type semiconductor

Fig. 2.27 A thin slab of n-type semiconductor with uniform illumination.

ionization. If the light is switched off at t = 0, the recombination rate exceeds the ionization rate, and the excess concentration decays exponentially to zero. It can be shown that, for t > 0, the concentration of holes is

$$p_n = p_{no} + p_{n'} e^{-t/\tau_h} , \qquad (2.64)$$

where τ_h is known as the lifetime of holes in the n-type semiconductor. The lifetime of minority carriers is an important parameter of semiconductor materials.

The above reasoning can be repeated for a p-type semiconductor, and it leads to the concept of a lifetime τ_e for minority-carrier electrons in a p-type material.

2.3.6.5 The pn junction

A pn junction functions as a diode, with the current-voltage characteristic shown in Fig. 2.28. The pn junction allows current to go from the p-side to the n-side (I > 0, V > 0, forward bias) and almost no current (negligibly small) from the n-side to the p-side (I < 0 but almost zero when V < 0, reverse bias). This effect is known as rectification, which is useful for diodes. When the applied voltage V is very negative, a large negative current flows, and this is known as breakdown.

Consider a pn junction at V = 0 (open circuited). Because the hole concentration is much higher in the p-side than the n-side, holes diffuse from the p-side to the n-side, thus causing the exposure of some acceptor anions near the junction in the p-side (Fig. 2.29). Similarly, because the conduction electron concentration is much higher in the n-side than the p-side, the conduction electrons very few carriers near the junction. This region is called the depletion region (or the space-charge layer,

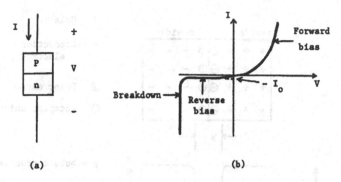

Fig. 2.28 The pn junction (a) configuration and (b) current (I) – voltage (V) relationship.

or the dipole layer, or the transition region). The exposed ions in the depletion region causes an electric field in the depletion region, such that the electric potential is higher in the n-side than the p-side. The difference in electric potential between the two sides is called the contact potential (V_o). Note that the contact potential is present even when the applied voltage V is zero. The contact potential is a barrier for the diffusion of holes from the p-side to the n-side, because holes want to go down in potential (i.e., more negative potential). Similarly, the contact potential is a barrier for the diffusion of conduction electrons from the n-side to the p-side, because electrons want to go up in potential (i.e., more positive potential). Both the diffusion of holes from the p-side to the n-side and the diffusion of electrons from the n-side to the p-side contribute to the diffusion current I_d, which flows from the p-side to the n-side.

There is a small concentration of holes in the n-side. When these holes approach the depletion region, they spontaneously go down the potential gradient and get to the p-side. Similarly, there is a small concentration of conduction electrons in the p-side. When these conduction electrons approach the depletion region, they spontaneously go up the potential gradient and get to the n-side. Such movements of the minority carriers constitute a drift current (I_o), which flows from the n-side to the p-side, i.e. in a direction opposite to that of the diffusion current I_d.

When the applied voltage V is zero, the current I is also zero, since there is an open circuit. Under this situation, $I_d = I_o$.

When the applied voltage V is negative (reverse bias), the applied potential is more positive in the n-side than the p-side. This causes the contact potential to increase to $V_o - V$ (since $V < 0$). As a result, the barrier to the diffusion current is increased, thus greatly lowering I_d, to the extent that $I_d \ll I_o$. Hence,

$$I = I_d - I_o \cong - I_o .$$

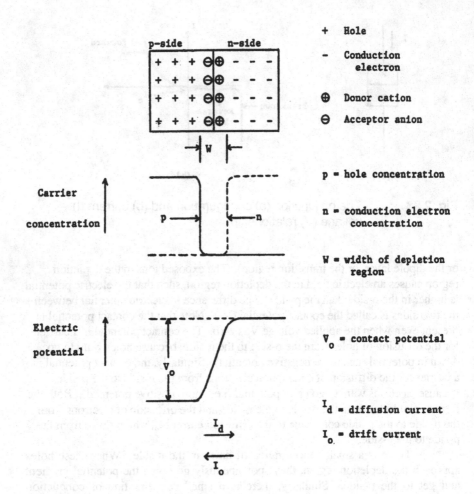

Fig. 2.29   A pn junction at bias voltage V = 0.

Because I_o is due to the minority carriers, it is bound to be small. Therefore, under reverse bias, a very small current flows from the n-side to the p-side.

When the applied voltage V is very negative, the contact potential becomes very large, resulting in an intense electric field in the depletion region. This strong electric field may tear electrons out of the covalent bonds, creating many more holes and conduction electrons. The holes and electrons get swept across the depletion region with high kinetic energies, knocking more valence electrons out of the covalent bonds. The consequence is a large reverse current (I <

0, V < 0). This phenomenon that occurs when V is sufficiently negative is known as breakdown.

When the applied voltage V is positive (forward bias), the applied potential is more positive in the p-side than the n-side, so the contact potential decreases to V_o - V. Hence, the barrier to the diffusion current is lowered and $I_d \gg I_o$. Therefore,

$$I = I_d - I_o \approx I_d.$$

In fact, I_d increases exponentially with increasing V.

Rectifying junctions are useful as diodes for electric circuits and can be used in combination with other junctions to provide transistors. They are based on semiconductors, most commonly silicon, as the doping of a semiconductor gives n-type and p-type materials for making pn-junctions.

In sharp contrast to the semiconductors widely used for rectifying junctions, cement-based materials can be used to make such junctions. Cement is a low-cost, mechanically rugged and electrically conducting material which can be rendered n-type or p-type by the use of appropriate admixtures, such as short carbon fibers for attaining p-type cement and short steel fibers for attaining n-type cement (cement itself is weakly n-type in relation to electronic conduction). The fibers also improve the structural properties, such as increasing the flexural strength and toughness and decreasing the drying shrinkage. Furthermore, cement-based junctions can be easily made by pouring the dissimilar cement mixes side by side. In addition, since cement is a structural material, cement-based junctions can be parts of a concrete structure, thereby allowing the structure to provide the rectifying function. This makes the structure multifunctional and smart.

Compared to semiconductor-based junctions, cement-based junctions involve negligible materials and processing costs, as single crystals, thin films, vacuum processing and clean rooms are not required for making cement-based junctions but are usually required for making semiconductor-based junctions. Moreover, cement-based junctions are mechanically rugged compared to semiconductor-based junctions. However, cement-based junctions are high in electrical resistance compared to semiconductor junctions.

As structural materials, cement-based materials have received much attention in terms of the mechanical properties, but relatively little attention in terms of the electrical conduction properties. Nevertheless, the electrical properties are relevant to the use of the structural materials for non-structural functions, such as sensing and electromagnetic interference (EMI) shielding. The ability to provide non-structural functions allows a structural material to be multifunctional, thus saving cost and enhancing durability. Furthermore, the electrical conduction properties shed light on the structure of the materials, particularly concerning the interfaces in the composite materials. Therefore, for both technological and fundamental reasons, the electrical conduction properties of cement-matrix composites are of interest.

2.4   Resistive behavior

Cement paste is electrically conductive, with DC resistivity at 28 days of curing around 5×10^3 Ω.m at room temperature. The resistivity is increased slightly (to 6×10^3 Ω.m) by the addition of silica fume (SiO$_2$ particles around 0.1 μm in size, in the amount of 15% by weight of cement), and increased more (to 7×10^3 Ω.m) by addition of latex (20% by weight of cement), which is a styrene-butadiene copolymer in the form of particles of size around 0.2 μm [3,4]. The higher the latex content, the higher the resistivity [5]. In the case of mortars (with fine aggregate, i.e., sand), the transition zone between the cement paste and the aggregate enhances the conductivity [6]. Whether aggregates (sand and stones) are present or not, the AC impedance spectroscopy technique for characterizing the frequency-dependent electrical behavior is useful for studying the microstructure [6-16].

The admixture effects mentioned above are small compared to the effect of adding short (5 mm) carbon fibers (0.5% by weight of cement), which decrease the resistivity to 2×10^2 Ω.m in the presence of silica fume and to 1×10^3 Ω.m in the presence of latex [3,4]. The lower resistivity when silica fume is present along with the carbon fibers than when latex is present is due to the greater degree of fiber dispersion provided by silica fume [17]. The fiber content mentioned above corresponds to 0.48 vol. % in the presence of silica fume and to 0.41 vol % in the presence of latex. These volume fractions are below the percolation threshold [18,19]. (Percolation refers to the situation in which there is a continuous electrical conduction path formed by the fibers, due to the contact between adjacent fibers in the composite.) Increasing the fiber content to 1.0% by weight of cement decreases the resistivity much more – to 8 Ω.m in the presence of silica fume and to 20 Ω.m in the presence of latex [3,4].

Fig. 2.30 [18] gives the volume electrical resistivity of carbon fiber composites at 7 days of curing. The resistivity decreases much with increasing fiber volume fraction, whether a second filler (silica fume or sand) is present or not. When sand is absent, the addition of silica fume decreases the resistivity at all carbon fiber volume fractions except the highest volume fraction of 4.24%; the decrease is most significant at the lowest fiber volume fraction of 0.53%. When sand is present, the addition of silica fume similarly decreases the resistivity, such that the decrease is most significant at fiber volume fractions below 1%. When silica fume is absent, the addition of sand decreases the resistivity only when the fiber volume fraction is below about 0.5%; at high fiber volume fractions, the addition of sand even increases the resistivity due to the porosity induced by the sand. Thus, the addition of a second filler (silica fume or sand) that is essentially non-conducting decreases the resistivity of the composite only at low volume fractions of the carbon fibers and the maximum fiber volume fraction for the resistivity to decrease is larger when the particle size of the filler is smaller. The resistivity decrease is attributed to the improved

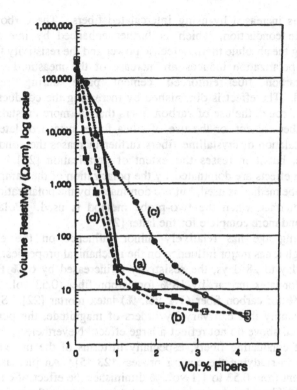

Fig. 2.30   Variation of the volume electrical resistivity with carbon fiber volume fraction.
(a)   Without sand, with methylcellulose, without silica fume.
(b)   Without sand, with methylcellulose, with silica fume.
(c)   With sand, with methylcellulose, without silica fume.
(d)   With sand, with methylcellulose, with silica fume.

fiber dispersion due to the presence of the second filler. Consistent with the improved fiber dispersion is the increased flexural toughness and strength due to the presence of the second filler [17].

The use of both silica fume and sand results in an electrical resistivity of 3.19 x 10^3 Ω.cm at a carbon fiber volume fraction of just 0.24 vol.%. This is an outstandingly low resistivity value compared to those of polymer-matrix composites with discontinuous conducting fillers at similar volume fractions.

Electrical conduction in cement reinforced by short carbon fibers below the percolation threshold is governed by carrier hopping across the fiber-matrix interface. The activation energy is decreased by increasing the fiber crystallinity, but is increased by using intercalated fibers. The carbon fibers contribute to hole conduction, which is further enhanced by intercalation, thereby decreasing the absolute thermoelectric power and the resistivity [20].

Electric polarization induces an increase of the measured electrical resistivity of carbon fiber-reinforced cement paste during resistivity measurement [21]. The effect is diminished by increasing the conductivity of the cement paste through the use of carbon fibers that are more crystalline, the increase of the fiber content, or the use of silica fume instead of latex as an admixture. Intercalation of crystalline fibers further increases the conductivity of the composite, but it increases the extent of polarization [20]. Voltage polarity switching effects are dominated by the polarization of the sample itself when the four-probe method is used, but are dominated by the polarization at the contact-sample interface when the two-probe method is used. Polarization reversal is faster and more complete for the latter [21].

The curing age has relatively minor influence on the electrical resistivity, although it has major influence on the mechanical properties. From a curing age of 1 day to 28 days, the resistivity is increased by 63% for plain mortar, by 18% for latex mortar, by 18% for carbon fiber (0.53 vol. %) latex mortar, and by 4% for carbon fiber (1.1 vol. %) latex mortar [22]. Since the resistivity is a quantity that can vary by orders of magnitude, the percentage increases mentioned above do not reflect a large effect. Nevertheless, the effect in the absence of conducting fibers, especially in terms of the impedance, is sufficient for use in studying the curing process [23-45]. An increase in the carbon fiber content from 0.53 to 1.1 vol. % diminishes the effect of curing age significantly, as the fibers become more dominant in governing the resistivity as the fiber content increases. The addition of latex also diminishes the effect of curing age.

Short steel fibers used as an admixture in cement paste also decrease the resistivity, such that the resistivity is lower in the presence of silica fume than in the presence of latex [46]. The contact electrical resistivity between stainless steel fiber and cement paste is around 6×10^2 $\Omega.m^2$ and is smaller if the fiber has been acid washed [47].

Fundamentally, the main difference between carbon fibers and steel fibers is that carbon fibers contribute to hole conduction whereas steel fibers contribute to electron conduction [4,46], as indicated by thermoelectric power measurement. Another difference is that the contact resistivity decreases with increasing bond strength for carbon fibers, due to the effect of interfacials voids, whereas the contact resistivity increases with increasing bond strength for steel fibers, due to the effect of the interfacial oxide phase [47,48].

The interface between steel fiber and cement matrix behaves similarly to that between steel reinforcing bar (rebar) and concrete. The latter is more common in practice than the former. The contact resistivity of the latter interface is around 6×10^3 $\Omega.m^2$ [49].

The insulating behavior of cement-matrix composites is relevant to the use as substrates for electronic packaging and as high voltage insulation. The attainment of a high resistivity requires thorough drying, as moisture decreases the resistivity [50,51]. Chloride and sulphate contamination also decrease the resistivity [52]. A resistivity as high as 10^{10} Ω.m has been attained [50-54].

2.5   Effect of temperature on resistive behavior

The dependence of the resistivity on temperature provides information on the energetics and mechanism of electrical conduction. Furthermore, this dependence allows the cement-matrix composite to serve as a thermistor, which is a thermometric device consisting of a material (typically a semiconductor, but in this case a cement-matrix composite) whose electrical resistivity decreases with rise in temperature.

Figure 2.31 [3] shows the current-voltage characteristic of carbon fiber (0.5% by weight of cement) silica fume (15% by weight of cement) cement paste at 38°C during stepped heating. The characteristic is linear below 5 V and deviates positively from linearity beyond 5 V. The deviation from linearity is due to Joule heating. The resistivity is obtained from the slope of the linear portion. The voltage at which the characteristic starts to deviate from linearity is referred to as the critical voltage.

Figure 2.32 shows a plot of the resistivity vs. temperature during heating and cooling for carbon fiber silica fume cement paste. The resistivity decreases upon heating and the effect is quite reversible upon cooling. That the

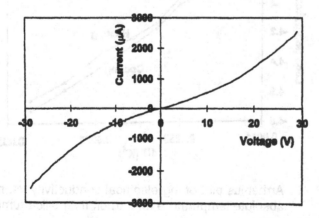

Fig. 2.31   Current-voltage characteristic of carbon fiber silica fume cement paste at 38°C during stepped heating.

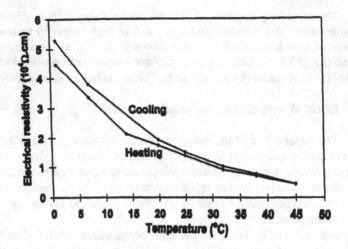

Fig. 2.32   Plot of volume electrical resistivity vs. temperature during heating and cooling for carbon fiber silica fume cement paste.

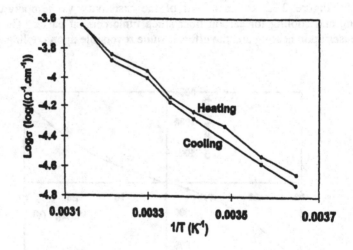

Fig. 2.33   Arrhenius plot of log electrical conductivity vs. reciprocal absolute temperature for carbon fiber silica fume cement paste.

resistivity is slightly increased after a heating-cooling cycle is probably due to thermal degradation of the material.

Fig. 2.33 shows the Arrhenius plot of log conductivity (conductivity = 1/resistivity) vs. reciprocal absolute temperature. The slope of the plot gives the activation energy, which is 0.390 ± 0.014 and 0.412 ± 0.017 eV during heating and cooling respectively. This energy corresponds to the barrier for electron motion across the interface between carbon fiber and cement paste [3]. The contact resistivity of this interface is around 50 $\Omega.m^2$, as obtained by single fiber electromechanical pull-out testing [48]. The contact resistivity is dependent on the fiber surface treatment and is higher when latex is present in the cement paste [48].

Results similar to those of carbon fiber silica fume cement paste were obtained with carbon fiber (0.5% by weight of cement) latex (20% by weight of cement) cement paste, silica fume cement paste, latex cement paste and plain cement paste. However, for all these four types of cement paste, (i) the resistivity is higher by about an order of magnitude, and (ii) the activation energy is lower by about an order of magnitude, as shown in Table 2.2. The critical voltage is higher when fibers are absent (Table 2.2).

2.6   Effect of strain on resistive behavior

Piezoresistivity (to be distinguished from piezoelectricity) is a phenomenon in which the electrical resistivity of a material changes with strain, which relates to stress. This phenomenon allows a material to serve as a strain/stress sensor. Applications of the stress/strain sensors include pressure sensors for aircraft and automobile components, vibration sensors for civil structures such as bridges and weighing-in-motion sensors for highways (weighing of vehicles). The first category tends to involve small sensors (e.g.,

Table 2.2   Resistivity, critical voltage and activation energy of five types of cement paste.

Formulation	Resistivity at 20°C (Ω.m)	Critical voltage at 20°C (V)	Activation energy (eV)	
			Heating	Cooling
Plain	$(4.87 \pm 0.37) \times 10^3$	10.80 ± 0.45	0.040 ± 0.006	0.122 ± 0.006
Silica fume	$(6.12 \pm 0.15) \times 10^3$	11.60 ± 0.37	0.035 ± 0.003	0.084 ± 0.004
Carbon fibers + silica fume	$(1.73 \pm 0.08) \times 10^2$	8.15 ± 0.34	0.390 ± 0.014	0.412 ± 0.017
Latex	$(6.99 \pm 0.12) \times 10^3$	11.80 ± 0.31	0.017 ± 0.001	0.025 ± 0.002
Carbon fibers + latex	$(9.64 \pm 0.08) \times 10^2$	8.76 ± 0.35	0.018 ± 0.001	0.027 ± 0.002

in the form of cement paste or mortar), and they will compete with silicon pressure sensors. The second and third categories tend to involve large sensors (e.g., in the form of precast concrete or mortar), and they will compete with silicon and acoustic, inductive and pneumatic sensors.

Piezoresistivity studies have been conducted mostly on polymer-matrix composites with fillers that are electrically conducting. These composite piezoresistive sensors work because strain changes the proximity between the conducting filler units, thus affecting the electrical resistivity. Tension increases the distance between the filler units, thus increasing the resistivity; compression decreases this distance, thus decreasing the resistivity.

Piezoresistivity in a structural material, such as a continuous fiber polymer-matrix composite, is particularly attractive, since the structural material becomes an intrinsically smart material that senses its own strain without the need for embedded or attached strain sensors. Not needing embedded or attached sensors means lower cost, greater durability, larger sensing volume (with the whole structure being able to sense) and absence of mechanical property degradation (which occurs in the case of embedded sensors).

The electrical resistance of a carbon fiber cement-matrix composite changes reversibly with strain, such that the gage factor (fractional change in resistance per unit strain) is up to 700 under compression or tension [56-70]. Both resistance (DC) and reactance (AC) increase reversibly upon tension and decreases reversibly upon compression, due to fiber pull-out upon microcrack opening (< 1 μm) and the consequent increase in fiber-matrix contact resistivity. The concrete contains as low as 0.2 vol.% short carbon fibers, which are preferably those that have been surface treated. The fibers do not need to touch one another in the composite. The treatment improves the wettability with water. The presence of a large aggregate decreases the gage factor, but the strain-sensing ability remains sufficient for practical use. Strain-sensing concrete works even when data acquisition is wireless. The applications include structural vibration control and traffic monitoring.

Figures 2.34 and 2.35 [66] show the fractional changes in the longitudinal and transverse resistivities respectively for carbon fiber silica fume cement paste at 28 days of curing during repeated uniaxial tensile loading at increasing strain amplitudes. The strain essentially returns to zero at the end of each cycle, indicating elastic deformation. The longitudinal strain is positive (i.e., elongation); the transverse strain is negative (i.e., shrinkage due to the Poisson Effect). Both longitudinal and transverse resistivities increase reversibly upon uniaxial tension. The reversibility of both strain and resistivity is more complete in the longitudinal direction than the transverse direction. The gage factor is 89 and −59 for the longitudinal and transverse resistances respectively.

Figures 2.36 and 2.37 [66] show corresponding results for silica-fume cement paste (without fiber). The strain is essentially totally reversible in both

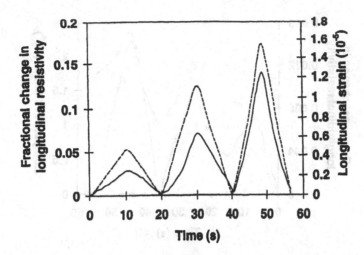

Fig. 2.34   Variation of the fractional change in longitudinal electrical resistivity with time (solid curve) and of the strain with time (dashed curve) during dynamic uniaxial tensile loading at increasing stress amplitudes within the elastic regime for carbon fiber silica fume cement paste.

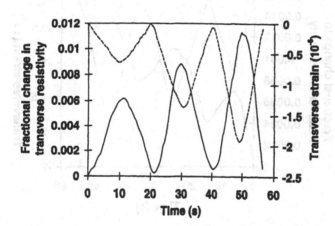

Fig. 2.35   Variation of the fractional change in transverse electrical resistivity with time (solid curve) and of the strain with time (dashed curve) during dynamic uniaxial tensile loading at increasing stress amplitudes within the elastic regime for carbon fiber silica fume cement paste.

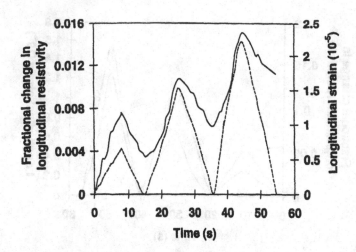

Fig. 2.36   Variation of the fractional change in longitudinal electrical resistivity with time (solid curve) and of the strain with time (dashed curve) during dynamic uniaxial tensile loading at increasing stress amplitudes within the elastic regime for silica fume cement paste.

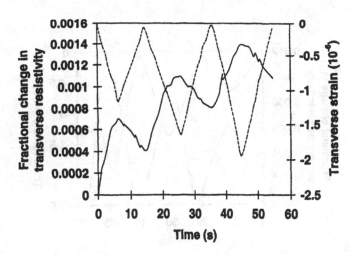

Fig. 2.37   Variation of the fractional change in transverse electrical resistivity with time (solid curve) and of the strain with time (dashed curve) during dynamic uniaxial tensile loading at increasing stress amplitudes within the elastic regime for silica fume cement paste.

the longitudinal and transverse directions, but the resistivity is only partly reversible in both directions, in contrast to the reversibility of the resistivity when fibers are present (Figs. 2.34 and 2.35). As in the case with fibers, both longitudinal and transverse resistivities increase upon uniaxial tension. However, the gage factor is only 7.2 and −7.1 for Figs. 2.36 and 2.37 respectively.

Comparison of Figs. 2.34-2.35 (with fibers) with Figs. 2.36-2.37 (without fibers) shows that fibers greatly enhance the magnitude, and the increase in both longitudinal and transverse resistivities upon uniaxial tension for cement pastes, whether with or without fibers, is attributed to defect generation [66]. In the presence of fibers, fiber bridging across microcracks occurs and slight fiber pull-out occurs upon tension, thus enhancing the possibility of microcrack closing and causing more reversibility in the resistivity change. The fibers are much more electrically conductive than the cement matrix. The presence of the fibers introduces interfaces between fibers and matrix. The degradation of the fiber-matrix interface due to fiber pull-out or other mechanisms is an additional type of defect generation which will increase the resistivity of the composite. Therefore, the presence of fibers greatly increases the gage factor.

The transverse resistivity increases upon uniaxial tension, even though the Poisson effect causes the transverse strain to be negative. This means that the effect of the transverse resistivity increase overshadows the effect of the transverse shrinkage. The resistivity increase is a consequence of the uniaxial tension. In contrast, under uniaxial compression, the resistance in the stress direction decreases at 28 days of curing. Hence, the effects of uniaxial tension on the transverse resistivity and of uniaxial compression on the longitudinal resistivity are different; the gage factors are negative and positive for these cases respectively.

The similarity of the resistivity change in longitudinal and transverse directions under uniaxial tension suggests a similarity in other directions as well. This means that the resistance can be measured in any direction in order to sense the occurrence of tensile loading. Although the gage factor is comparable in both longitudinal and transverse directions, the fractional change in resistance under uniaxial tension is much higher in the longitudinal direction than the transverse direction. Thus, the use of the longitudinal resistance for practical self-sensing is preferred.

Piezoresistivity also occurs in cement-matrix composites with continuous carbon fibers [71]. The electrical resistance in the fiber direction, as measured using surface electrical contacts, increases upon tension in the same direction. The resistance increase is mostly reversible, such that the irreversible portion increases with the stress amplitude. The effect is attributed to fiber-matrix interface degradation, which is partly irreversible. The gage factor is up to 60.

2.7   Effect of damage on resistive behavior

Concrete, with or without admixtures, is capable of sensing major and minor damage – even damage during elastic deformation – due to the electrical resistivity increase that accompanies damage [57,61,66,72]. That both strain and damage can be sensed simultaneously through resistance measurement means that the strain/stress condition (during dynamic loading) under which damage occurs can be obtained, thus facilitating damage origin identification. Damage is indicated by a resistance increase, which is larger and less reversible when the stress amplitude is higher. The resistance increase can be a sudden increase during loading. It can also be a gradual shift of the baseline resistance.

Fig. 2.38 [66] shows the fractional change in resistivity along the stress axis as well as the strain during repeated compressive loading at an increasing stress amplitude for plain cement paste at 28 days of curing. The strain varies linearly with the stress up to the highest stress amplitude. The strain returns to zero at the end of each cycle of loading. During the first loading, the resistivity increases due to damage initiation. During the subsequent unloading, the resistivity continues to increase, probably due to opening of the microcracks generated during loading. During the second loading, the resistivity decreases slightly as the stress increases up to the maximum stress of the first cycle

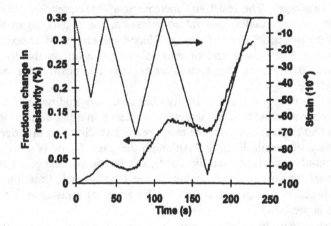

Fig. 2.38   Variation of the fractional change in electrical resistivity with time and of the strain (negative for compressive strain) with time during dynamic compressive loading at increasing stress amplitudes within the elastic regime for plain cement paste at 28 days of curing.

(probably due to closing of the microcracks) and then increases as the stress increases beyond this value (probably due to the generation of additional microcracks). During unloading in the second cycle, the resistivity increases significantly (probably due to opening of the microcracks). During the third loading, the resistivity essentially does not change (or decreases very slightly) as

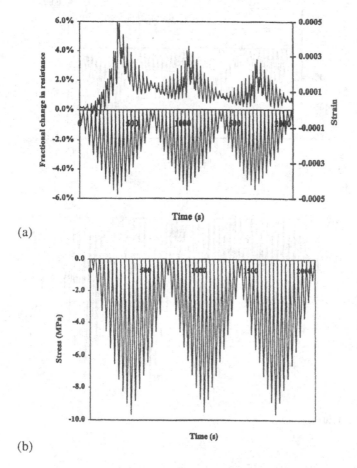

(a)

(b)

Fig. 2.39   Fractional change in resistance (a), strain (a) and stress (b) during repeated compressive loading at increasing and decreasing stress amplitudes, the highest of which was 60% of the compressive strength, for carbon fiber concrete at 28 days of curing.

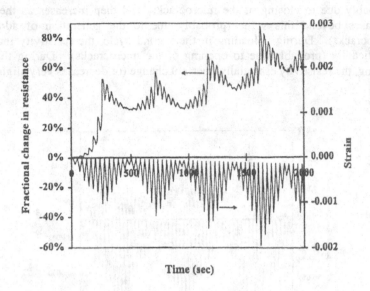

(a)

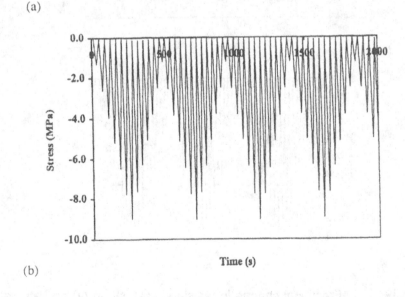

(b)

Fig. 2.40   Fractional change in resistance (a), strain (a) and stress
                  (b) during repeated compressive loading at increasing
                  and decreasing stress amplitudes, the highest of which
                  was >90% of the compressive strength, for carbon fiber
                  concrete at 28 days of curing.

the stress increases to the maximum stress of the third cycle (probably due to the balance between microcrack generation and microcrack closing). Subsequent unloading causes the resistivity to increase very significantly (probably due to opening of the microcracks).

Fig. 2.39 [72] shows the fractional change in resistance, strain and stress during repeated compressive loading at increasing and decreasing stress amplitudes for carbon fiber (0.18 vol.%) concrete (with fine and coarse aggregates) at 28 days of curing. The highest stress amplitude is 60% of the compressive strength. A group of cycles in which the stress amplitude increases cycle by cycle and then decreases cycle by cycle back to the initial low stress amplitude is hereby referred to as a group. Fig. 2.39 shows the results for three groups. The strain returns to zero at the end of each cycle for any of the stress amplitudes, indicating elastic behavior. The resistance decreases upon loading in each cycle, as in Fig. 2.33. An extra peak at the maximum stress of a cycle grows as the stress amplitude increases, resulting in two peaks per cycle. The original peak (strain induced) occurs at zero stress, while the extra peak (damage induced) occurs at the maximum stress. Hence, during loading from zero stress within a cycle, the resistance drops and then increases sharply, reaching the maximum resistance of the extra peak at the maximum stress of the cycle. Upon subsequent unloading, the resistance decreases and then increases as unloading continues, reaching the maximum resistance of the original peak at zero stress. In the part of this group where the stress amplitude decreases cycle by cycle, the extra peak diminishes and disappears, leaving the original peak as the sole peak. In the part of the second group where the stress amplitude increases cycle by cycle, the original peak (peak at zero stress) is the sole peak, except that the extra peak (peak at the maximum stress) returns in a minor way (more minor than in the first group) as the stress amplitude increases. The extra peak grows as the stress amplitude increases, but, in the part of the second group in which the stress amplitude decreases cycle by cycle, it quickly diminishes and vanishes, as in the first group. Within each group, the amplitude of resistance variation increases as the stress amplitude increases and decreases as the stress amplitude subsequently decreases.

The greater the stress amplitude, the larger and the less reversible is the damage-induced resistance increase (the extra peak). If the stress amplitude has been experienced before, the damage-induced resistance increase (the extra peak) is small, as shown by comparing the result of the second group with that of the first group (Fig. 2.39), unless the extent of damage is large (Fig. 2.40 [72] for a highest stress amplitude of >90% the compressive strength). When the damage is extensive (as shown by a modulus decrease), damage-induced resistance increase occurs in every cycle, even at a decreasing stress amplitude, and it can overshadow the strain-induced resistance decrease (Fig. 2.40). Hence, the damage-induced resistance increase occurs mainly during loading (even within the elastic regime), particularly at a stress above that in prior cycles, unless the stress amplitude is high and/or damage is extensive.

At a high stress amplitude, the cycle-by-cycle damage-induced resistance increase as the stress amplitude increases causes the baseline resistance to increase irreversibly (Fig. 2.40). The baseline resistance in the regime of major damage (with a decrease in modulus) provides a measure of the extent of damage (i.e., condition monitoring). This measure works in the loaded or unloaded state. In contrast, the measure using the damage-induced resistance increase (Fig. 2.39) works only during stress increase and indicates the occurrence of damage (whether minor or major) as well as the extent of damage.

2.8   Thermoelectric behavior

The Seebeck effect [4,46,73-75] is a thermoelectric effect (i.e., an effect involving the conversion between thermal energy and electrical energy) which is the basis for thermocouples for temperature measurement. It also allows the conversion from thermal energy to electrical energy. This effect involves charge carriers moving from a hot point to a cold point within a material, thereby resulting in a voltage difference between the two points. The Seebeck coefficient is the voltage difference (hot minus cold) per unit temperature difference (hot minus cold) between the two points. Negative carriers (electrons) make it more negative, and positive carriers (holes) make it more positive.

The Seebeck effect in carbon fiber-reinforced cement paste involves electrons from the cement matrix [4] and holes from the fibers [73-75], such that the two contributions are equal at the percolation threshold, a fiber content between 0.5% and 1.0% by weight of cement [4]. The hole contribution increases monotonically with increasing fiber content below and above the percolation threshold [4]. It also increases with intercalation of the carbon fibers with an acceptor such as bromine [20], so that the absolute thermoelectric power reaches +17 μV/°C (Table 2.3).

Cement containing stainless steel fibers (60 μm diameter) is even more negative (as negative as −69 μV/°C, Table 2.3) in the absolute thermoelectric power than cement without fiber [46]. The attainment of a very negative thermoelectric power is attractive, since a material with a positive thermoelectric power and a material with negative thermoelectric power are two very dissimilar materials, the junction of which is a thermocouple junction. (The greater the dissimilarity, the more sensitive is the thermocouple.)   A cement-based thermocouple and electric current rectification have been attained by using a cement-based pn junction [76].

Table 2.3 and Fig. 2.41 show the thermopower results. The absolute thermoelectric power is much more negative for all the steel fiber cement pastes compared to all the carbon fiber cement pastes. An increase of the steel fiber content from 0.5% to 1.0% by weight of cement makes the absolute thermoelectric power more negative, whether silica fume (or latex) is present or not. An increase of the steel fiber content also increases the reversibility and linearity of the change in Seebeck voltage with the temperature difference between the hot and cold ends.

Table 2.3 Volume electical resistivity and absolute thermoelectric power of various cement pastes with steel fibers (S_f), pristine carbon fibers (C_f) or intercalated carbon fibers (C_f').

Cement paste	Volume fraction fibers	Resistivity ($\Omega.m$)	Absolute thermoelectric power ($\mu V/°C$)
Plain	0	$(4.7 \pm 0.4) \times 10^3$	-1.99 ± 0.03
SF	0	$(5.8 \pm 0.4) \times 10^3$	-2.03 ± 0.02
L	0	$(6.8 \pm 0.6) \times 10^3$	-2.06 ± 0.02
$S_f (0.5^*)$	0.10%	$(7.8 \pm 0.5) \times 10^2$	-53.3 ± 4.8
$S_f (1.0^*)$	0.20%	$(4.8 \pm 0.4) \times 10^2$	-59.1 ± 5.2
$S_f (0.5^*) + SF$	0.10%	$(5.6 \pm 0.5) \times 10^2$	-57.1 ± 3.9
$S_f (1.0^*) + SF$	0.20%	$(3.2 \pm 0.3) \times 10^2$	-68.5 ± 4.5
$S_f (0.5^*) + L$	0.085%	$(1.4 \pm 0.1) \times 10^3$	-50.4 ± 3.2
$S_f (1.0^*) + L$	0.17%	$(1.1 \pm 0.1) \times 10^3$	-57.7 ± 5.0
$C_f (0.5^*) + SF$	0.48%	$(1.5 \pm 0.1) \times 10^2$	-0.89 ± 0.09
$C_f (1.0^*) + SF$	0.95%	8.3 ± 0.5	$+0.48 \pm 0.11$
$C_f (0.5^*) + L$	0.41%	$(9.7 \pm 0.6) \times 10^2$	-1.14 ± 0.05
$C_f (1.0^*) + L$	0.82%	$(1.8 \pm 0.2) \times 10^1$	-0.24 ± 0.08
$C_f' (0.5^*) + SF$	0.4%	/	$+11.5 \pm 1.13$
$C_f' (1.0^*) + SF$	0.8%	/	$+16.6 \pm 1.32$
$C_f' (0.5^*) + L$	0.3%	/	$+7.42 \pm 1.09$
$C_f' (1.0^*) + L$	0.7%	/	$+10.2 \pm 1.07$

* % by weight of cement
SF: silica fume
L: latex

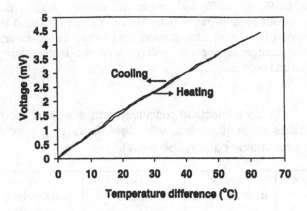

Fig. 2.41   Variation of the Seebeck voltage (with copper as the reference) vs. the temperature difference during heating and cooling for steel fiber silica fume cement paste containing steel fibers in the amount of 1.0% by weight of cement.

Table 2.3 shows that the volume electrical resistivity is much higher for the steel fiber cement pastes than the corresponding carbon fiber cement pastes. This is attributed to the much lower volume fraction of fibers in the former (Table 2.3). An increase in the steel or carbon fiber content from 0.5% to 1.0% by weight of cement decreases the resistivity, though the decrease is more significant for the carbon fiber case than the steel fiber case. That the resistivity decrease is not large when the steel fiber content is increased from 0.5% to 1.0% by weight of cement and that the resistivity is still high at a steel fiber content of 1.0% by weight of cement suggest that a steel fiber content of 1.0% by weight of cement is below the percolation threshold.

Whether with or without silica fume (or latex), the change of the Seebeck voltage with temperature is more reversible and linear at a steel fiber content of 1.0% by weight of cement than at a steel fiber content of 0.5% by weight of cement. This is attributed to the larger role of the cement matrix at the lower steel fiber content and the contribution of the cement matrix to the irreversibility and non-linearity. Irreversibility and non-linearity are particularly significant when the cement paste contains no fiber.

From the practical point of view, the steel fiber silica fume cement paste containing steel fibers in the amount of 1.0% by weight of cement is particularly attractive for use in temperature sensing, as the absolute thermoelectric power is the highest in magnitude (-68 μV/°C) and the variation of the Seebeck voltage with the temperature difference between the hot and cold ends is reversible and linear. The absolute thermoelectric power is as high as those of commercial thermocouple materials.

2.9 Electromagnetic behavior

Cement-matrix composites interact with electromagnetic radiation by reflection and absorption. The reflection component can be greatly enhanced by the use of an electrically conductive admixture, such as short carbon fibers.

Electromagnetic interference (EMI) shielding refers to the reflection and/or adsorption of electromagnetic radiation by a material, which thereby acts as a shield against the penetration of the radiation through the shield. As electromagnetic radiation, particularly that at high frequencies (e.g., radio waves, such as those emanating from cellular phones) tend to interfere with electronics (e.g., computers), EMI shielding of both electronics and radiation source is needed and is increasingly required by governments around the world. The importance of EMI shielding relates to the high demand of today's society on the reliability of electronics and the rapid growth of radio frequency radiation sources.

Electromagnetic radiation at high frequencies penetrates only the near surface region of an electrical conductor. This is known as the skin effect. Due to the skin effect, a composite material having a conductive filler with a small unit size (e.g., particle size, fiber diameter, etc.) of the filler is more effective than one having a conductive filler with a large unit size of the filler. For effective use of the entire cross section of a filler unit for shielding, the unit size of the filler should be comparable to or less than the skin depth. Therefore, a filler of unit size 1 μm or less is typically preferred, though such a small unit size is not commonly available for most fillers, and the dispersion of the filler is more difficult when the filler unit size decreases.

Cement is slightly conducting, so the use of a cement matrix also allows the conductive filler units in the composite to be electrically connected, even when the filler units do not touch one another. Thus, cement-matrix composites have higher shielding effectiveness than corresponding polymer-matrix composites in which the polymer matrix is insulating [77]. A shielding effectiveness of 40 dB at 1 GHz has been attained in a cement-matrix composite containing 1.5 vol. % discontinuous 0.1 μm-diameter carbon filaments [78], and an effectiveness of 70 dB at 1.5 GHz has been attained in a cement-matrix composite containing 0.72 vol.% 8 μm-diameter stainless steel fibers [79]. Moreover, cement is less expensive than polymers and cement-matrix composites are useful for the shielding of rooms in a building [78-81]. In addition, the reflectivity renders the ability to provide lateral guidance in the automatic highway technology, as a traffic lane with a radio wave reflecting concrete as an overlay in either the middle portion or the edge portions along the length of the lane reflects the wave emitted from a vehicle that is installed with a radio wave transmitter and receiver [77].

A disadvantage of EMI shielding is the inability of penetration of television signals. The combined use of carbon fibers and carbon beads alleviates this problem [82].

The interaction of electromagnetic radiation with concrete can also be utilized for nondestructive evaluation [83,84]. For this application, high reflectivity is not desirable.

2.10   Conclusion

The resistive, piezoresistive, thermoelectric and electromagnetic behavior of cement-based materials can be greatly modified by the use of admixtures, such as short carbon fibers and short steel fibers. Short carbon fibers are effective for enhancing the piezoresistive behavior and enhancing the effect of damage on the resistive behavior. Short carbon fibers are effective for enhancing the p-type Seebeck effect, whereas short steel fibers are effective for enhancing the n-type Seebeck effect. Submicron diameter carbon filaments are effective for enhancing the electromagnetic reflection behavior.

References

1.   S. Wen, S. Wang and D.D.L. Chung, *Sensors & Actuators* A 78, 180-188 (1999).
2.   S. Wen and D.D.L. Chung, *Cem. Concr. Res.* 29(6), 961-965 (1999).
3.   S. Wen and D.D.L. Chung, *Cem. Concr. Res.* 29(6), 961-965 (1999).
4.   S. Wen and D.D.L. Chung, *Cem. Concr. Res.* 29(12), 1989-1993 (1999).
5.   X. Fu and D.D.L. Chung, *Cem. Concr. Res.* 26(7), 985-991 (1996).
6.   P.J. Tumidajski, *Cem. Concr. Res.* 26(4), 529-534 (1996).
7.   G. Ping, X. Ping and J.J. Beaudoin, *Cem. Concr. Res.* 23(3), 581-591 (1993).
8.   T.O. Mason, S.J. Ford, J.D. Shane, J.-H. Hwang and D.D. Edwards, *Adv. Cem. Res.* 10(4), 143-150 (1998).
9.   D.E. MacPhee, D.C. Sinclair and S.L. Stubbs, *J. Mater. Sci. Letters* 15(18), 1566-1568 (1996).
10.   S.L. Cormack, D.E. MacPhee and D.C. Sinclair, *Adv. Cem. Res.* 10(4), 151-159 (1998).
11.   M. Keddam, H. Takenouti, X.R. Novoa, C. Andrade and C. Alonso, *Cem. Concr. Res.* 27(8), 1191-1201 (1997).
12.   W.J. McCarter and G. Starrs, *J. Mater. Sci. Letters* 16(8), 605-607 (1997).
13.   P. Gu and J.J. Beaudoin, *J. Mater Sci. Letters* 15(2), 182-184 (1996).

14. Z. Xu, P. Gu, P. Xie and J.J. Beaudoin, *Cem. Concr. Res.* 23(4), 853-862 (1993).
15. S.J. Ford, T.O. Mason, B.J. Christensen, R.T. Coverdale, H.M. Jennings and E.J. Garboczi, *J. Mater. Sci.* 30(5), 1217-1224 (1995).
16. P. Gu, Z. Xu, P. Xie and J.J. Beaudoin, *Cem. Concr. Res.* 23(3), 531-540 (1993).
17. P.-W. Chen, X. Fu and D.D.L. Chung, *ACI Mater. J.* 94(2), 147-155 (1997).
18. P. Chen and D.D.L. Chung, *J. Electron. Mater.* 24(1), 47-51 (1995).
19. Z. Shui, J. Li, F. Huang and D. Yang, *J. Wuhan Univ. Tech.* 10(4), 37-41 (1995).
20. S. Wen and D.D.L. Chung, *Carbon* 39, 369-373 (2001).
21. S. Wen and D.D.L. Chung, *Cem. Concr. Res.* 31(2), 141-147 (2001).
22. X. Fu and D.D.L. Chung, *Cem. Concr. Res.* 25(4), 689-694 (1995).
23. M.S. Morsy, *Cem. Concr. Res.* 29, 603-606 (1999).
24. J.G. Wilson and N.K. Gupta, *Building Res. Info.* 24(4), 209-212 (1996).
25. S.A. Abo El-Enein, M.F. Kotkata, G.B. Hanna, M. Saad and M.M. Abd El Razek, *Cem. Concr. Res.* 25(8), 1615-1620 (1995).
26. H.C. Kim, S.Y. Kim and S.S. Yoon, *J. Mater. Sci.* 30(15), 3768-3772 (1995).
27. C. Alonso, C. Andrade, M. Keddam, X.R. Novoa and H. Takenouti, *Mater. Sci. Forum* 289-292(pt 1), 15-28 (1998).
28. P. Gu, P. Xie, J.J. Beaudoin and R. Brousseau, *Cem. Concr. Res.* 23(1), 157-168 (1993).
29. P. Xie, P. Gu, Z. Xu and J.J. Beaudoin, *Cem. Concr. Res.* 23(2), 359-367 (1993).
30. P. Gu, P. Xie, Y. Fu and J.J. Beaudoin, *Cem. Concr. Res.* 24(1), 86-88 (1994).
31. P. Gu, P. Xie, Y. Fu and J.J. Beaudoin, *Cem. Concr. Res.* 24(1), 89-91 (1994).
32. P. Xie, P. Gu, Y. Fu and J.J. Beaudoin, *Cem. Concr. Res.* 24(1), 92-94 (1994).
33. P. Xie, P. Gu, Y. Fu and J.J. Beaudoin, *Cem. Concr. Res.* 24(4), 704-706 (1994).
34. P. Gu, P. Xie and J.J. Beaudoin, *Cem., Concr. & Aggregates* 17(1), 92-97 (1995).
35. W.J. McCarter, *Cem. Concr. Res.* 24(6), 1097-1110 (1994).
36. B.J. Christensen, R.T. Coverdale, R.A. Olson, S.J. Ford, E.J. Garboczi, H.M. Jennings and T.O. Mason, *J. American Ceramic Society* 77(11), 2789-2804 (1994).
37. P. Gu, P. Xie and J.J. Beaudoin, *Cem., Concr. & Aggregates* 17(2), 113-118 (1995).
38. P. Xie, P. Gu and J.J. Beaudoin, *J. Mater. Sci.* 31(1), 144-149 (1996).

39.   S.S. Yoon, H.C. Kim and R.M. Hill, *J. Physics D-Applied Physics* 29(3), 869-875 (1996).
40.   G.M. Moss, D.J. Christensen, T.O. Mason and H.M. Jennings, *Adv. Cem. Based Mater.* 4(2), 68-75 (1996).
41.   W.J. McCarter, *J. Mater. Sci.* 31(23), 6285-6292 (1996).
42.   S.J. Ford, J.-H. Hwang, J.D. Shane, R.A. Olson, G.M. Moss, H.M. Jennings and T.O. Mason, *Adv. Cem. Based Mater.* 5(2), 41-48 (1997).
43.   J.M. Torrents and R. Pallas-Areny, *Conf. Record – IEEE Instrumentation and Measurement Technology Conference,* vol 2, (1997), IEEE, Piscataway, NJ, 1089-1093.
44.   J.M. Torrents, J. Roncero and R. Gettu, *Cem. Concr. Res.* 28(9), 1325-1333 (1998).
45.   C. Andrade, V.M. Blanco, A. Collazo, M. Keddam, X.R. Novoa and H. Takenouti, *Electrochimica Acta* 44(24), 4313-4318 (1999).
46.   S. Wen and D.D.L. Chung, *Cem. Concr. Res.* 30(4), 661-664 (2000).
47.   X. Fu and D.D.L. Chung, *Composite Interfaces* 4(4), 197-211 (1997).
48.   X. Fu, W. Lu and D.D.L. Chung, *Cem. Concr. Res.* 26(7), 1007-1012 (1996).
49.   X. Fu and D.D.L. Chung, *Composite Interfaces* 6(2), 81-92 (1999).
50.   D.E. Wilkosz and J.F. Young, *J. Am. Ceram. Soc.* 78(6), 1673-1679 (1995).
51.   D. Buerchler, B. Elsener and H. Boehni, *Mat. Res. Soc. Symp. Proc.* 411, 407-412 (1996).
52.   M. Saleem, M. Shameem, S.E. Hussain and M. Maslehuddin, *Construction & Building Mater.* 10(3), 209-214 (1996).
53.   B.P. Borglum, J.F. Young and R.C. Buchanan, *Adv. Cem. Bas. Mat.* 1(1), 47-50 (1993).
54.   J.F. Young, *Mat. Tech.* 9(3/4), 63-67 (1994).
55.   W.J. McCarter, *J. American Ceramic Society* 78(2), 411-415 (1995).
56.   X. Fu, W. Lu and D.D.L. Chung, *Carbon* 36(9), 1337-1345 (1998).
57.   P. Chen and D.D.L. Chung, *Smart Mater. Struct.* 2, 22-30 (1993).
58.   P. Chen and D.D.L. Chung, *Composites*, Part B 27B, 11-23 (1996).
59.   P. Chen and D.D.L. Chung, *J. Am. Ceram. Soc.* 78(3), 816-818 (1995).
60.   D.D.L. Chung, *Smart Mater. Struct.* 4, 59-61 (1995).
61.   P. Chen and D.D.L. Chung, *ACI Mater. J.* 93(4), 341-350 (1996).
62.   X. Fu and D.D.L. Chung, *Cem. Concr. Res.* 26(1), 15-20 (1996).
63.   X. Fu, E. Ma, D.D.L. Chung and W.A. Anderson, *Cem. Concr. Res.* 27(6), 845-852 (1997).
64.   X. Fu and D.D.L. Chung, *Cem. Concr. Res.* 27(9), 1313-1318 (1997).
65.   X. Fu, W. Lu and D.D.L. Chung, *Cem. Concr. Res.* 28(2), 183-187 (1998).
66.   D.D.L. Chung, *TANSO* (190), 300-312 (1999).
67.   Z. Shi and D.D.L. Chung, *Cem. Concr. Res.* 29(3), 435-439 (1999).

68. Q. Mao, B. Zhao, D. Sheng and Z. Li, *J. Wuhan Univ. Tech.*, Mater. Sci. Ed. 11(3), 41-45 (1996).
69. Q. Mao, B. Zhao, D. Shen and Z. Li, *Fuhe Cailiao Xuebao/Acta Materiae Compositae Sinica* 13(4), 8-11 (1996).
70. M. Sun, Q. Mao and Z. Li, *J. Wuhan Univ. Tech.*, Mater. Sci. Ed. 13(4), 58-61 (1998).
71. S. Wen and D.D.L. Chung, *Cem. Concr. Res.* 29(3), 445-449 (1999).
72. D. Bontea, D.D.L. Chung and G.C. Lee, *Cem. Concr. Res.* 30(4), 651-659 (2000).
73. M. Sun, Z. Li, Q. Mao and D. Shen, *Cem. Concr. Res.* 29(5), 769-771 (1999).
74. M. Sun, Z. Li, Q. Mao and D. Shen, *Cem. Concr. Res.* 28(12), 1707-1712 (1998).
75. M. Sun, Z. Li, Q. Mao and D. Shen, *Cem. Concr. Res.* 28(4), 549-554 (1998).
76. S. Wen and D.D.L. Chung, *J. Mater. Res.* 16(7), 1989-1993 (2001).
77. X. Fu and D.D.L. Chung, *Carbon* 36(4), 459-462 (1998).
78. S. Wen and D.D.L. Chung, *Cem. Concr. Res.*, in press.
79. L. Gnecco, *Evaluation Eng.* 38(3), 3 pp (1999).
80. S.-S. Lin, *SAMPE J.* 30(5), 39-45 (1994).
81. Y. Kurosaki and R. Satake, *IEEE Int. Symp. Electromagnetic Compatibility*, (1994) IEEE, Piscataway, NJ, pp. 739-740.
82. Y. Shimizu, A. Nishikata, N. Maruyama and A. Sugiyama, *Nippon Terebijon Gakkaishi (J. Institute of Television Engineers of Japan)* 40(8), 780-785 (1986).
83. I.L. Al-Qadi, O.A. Hazim, W. Su and S.M. Riad, *J. Mater. Civil Eng.* 7(3), 192-198 (1995).
84. O. Buyukozturk and H.C. Rhim, *Cem. Concr. Res.* 25(5), 1011-1022 (1995).

3

Cement-Based Materials for Piezoresistivity (Strain Sensing)

3.1   Introduction

Strain sensing (related to stress sensing, but distinct from damage sensing) is relevant to structural vibration control, traffic monitoring and weighing (e.g., (i) the weighing of trucks that are moving, (ii) the weighing of all the people in each room of a building for the purpose of room occupancy monitoring and the use of the information for controlling lighting, ventilation, air-conditioning and heating (i.e., for energy saving), (iii) the detection of people inside or outside a building for enhancing building security, and (iv) the weighing of cargo).

Conventional applications of the stress/strain sensors include pressure sensors for aircraft and automobile components, vibration sensors for civil structures such as bridges and weighing-in-motion sensors for highways. The first category tends to involve small sensors (e.g., in the form of cement paste or mortar), and they will compete with silicon pressure sensors. The second and third categories tend to involve large sensors (e.g., in the form of precast concrete or mortar), and they will compete with silicon, acoustic, inductive and pneumatic sensors.

Cement reinforced with short carbon fibers is capable of sensing its own strain due to the effect of strain on the electrical resistivity (Figs. 2.32 and 2.33) [1-14]. As observed at 28 days of curing, the resistivity in the stress and transverse directions increases upon tension, due to slight fiber pull-out that accompanies crack opening, and decreases upon compression, due to slight fiber push-in that accompanies crack closing [6-11,13,14]. This electromechanical phenomenon, called piezoresistivity (i.e., change of the electrical resistivity with strain), allows the use of electrical resistance measurement (DC or AC) to monitor the strain of the cement-based material, which is itself the sensor. This means that the cement-based material is self-sensing. In contrast to the conventional method of using embedded or attached strain sensors [15,16], self-sensing involves low cost, high durability, large sensing volume and the absence of mechanical property degradation (which tends to occur in case of embedded sensors).

Piezoresistivity studies have been mostly conducted on polymer-matrix composites with fillers that are electrically conducting. These composite peizoresistive sensors work because strain changes the proximity between the conducting filler units, thus affecting the electrical resistivity. Tension increases the distance between the filler units, thus increasing the resistivity; compression decreases this distance, thus decreasing the resistivity.

Composite piezoresistive materials include polymer-matrix composites containing continuous carbon fibers [17-29], carbon black [30-32], metal particles [31], short carbon fibers [32,33], cement-matrix composites containing short carbon fibers [1,5-7,9,34], and ceramic-matrix composites containing silicon carbide whiskers [35]. The sensing of reversible strain had been observed in polymer-matrix and cement-matrix composites [1,5-7,9,17-21,30,31,33,34].

The presence of electrically conductive fibers in the cement-based material is necessary for the piezoresistivity to be sufficient in magnitude and in reversibility. In the absence of conductive fibers, the piezoresistivity is weak and has substantial irreversibility, if at all observable, as shown in the case of cement-based materials without fibers (Fig. 2.34 and 2.35) [35] and with non-conductive (polyethylene) short fibers [36]. Although conductive fibers are important for piezoresistivity, they are preferably discontinuous (around 5 mm in length, unless stated otherwise), due to the low cost of short fibers compared to continuous fibers and the amenability of short fibers for incorporation in the concrete mix by mixing, and are typically used at a volume fraction below the percolation threshold, which refers to the volume fraction above which the fibers touch one another to form a continuous electrical path. The fibers are not the sensors; they are an additive for rendering significant piezoresistivity to the cement-based material, which is the sensor. A low fraction of fibers is preferred for the purpose of maintaining low cost, high workability and high compressive strength.

Steel fibers are even more conductive than carbon fibers. Short steel fibers (typically of diameter 500 μm) are used in cement-based materials to enhance the tensile, flexural and shear properties [37-43] and the abrasion resistance [44], decrease the drying shrinkage [45], increase the effectiveness for electromagnetic interference (EMI) shielding [46] and provide controlled electrical resistivity [47]. Moreover, stainless steel fibers of diameter 60 μm render piezoresistivity to a cement-based material, as shown under compression, though the phenomenon is noisy in that the resistivity does not vary smoothly with the strain [36]. The large diameter of the steel fibers compared to carbon fibers (15 μm) was believed to be the cause of the inferior performance of the steel fiber cement-based material [36]. However, carbon fiber (15 μm diameter) cement paste is a better piezoresistive strain sensor than stainless steel fiber (8 μm diameter) cement paste at a similar fiber volume fraction, as shown by a higher signal-to-noise ratio and better reversibility upon unloading [48]. The difference in performance of carbon fiber cement and steel fiber cement is attributed to a difference in the piezoresistivity mechanism.

3.2   Effect of fiber type on the piezoresistive behavior

The experimental results presented in this section were all obtained at 28 days of curing, using Type I portland cement. For details on the materials processing, please refer to the literature cited.

3.2.1   Cement paste containing 0.72 vol.% short steel fibers

Fig. 3.1 [48] shows the variation of the fractional change in resistivity with strain and stress for cement paste containing 0.72 vol.% short steel fibers (8 µm diameter) under repeated tension. Both resistivity and strain increased with increasing stress with partial reversiblity. The higher the stress amplitude, the higher were both the strain and the resistivity.

Fig. 3.2 [48] shows corresponding results obtained under compression. The strain was mostly reversible, but the resistivity decrease upon compression was noisy and the resistivity showed an irreversible increase after each stress cycle.

3.2.2   Cement paste containing 0.36 vol.% short steel fibers

Figs. 3.3 [48] and 3.4 [48] show the piezoresistivity results for cement paste containing 0.36 vol.% short steel fibers (8 µm diameter) under tension and compression respectively. The resistivity increased upon tension and decreased upon compression, as observed for cement paste containing 0.72 vol.% steel fibers (Figs. 3.1 and 3.2). However, the resistivity change and strain were more reversible, both under tension and compression.

3.2.3   Cement paste containing 0.5 vol.% short carbon fibers

Figs. 3.5 and 3.6 [13,14] show the piezoresistivity results for cement paste containing 0.5 vol.% short carbon fibers (15 µm diameter) under tension and compression respectively. The strain was totally reversible and was linearly related to the stress. The resistivity increased with tensile strain and decreased with compressive strain, such that the effect was totally reversible, except for an irreversible increase at the end of the first compression cycle. The resistivity variation was much less noisy and much more reversible than that observed for the two steel fiber cement pastes (Fig. 3.1-3.4).

The use of four electrical contacts on the same surface of 0.35 vol.% short carbon fiber cement mortar under flexure has been shown to provide measurement of the change in surface resistivity on the tension side or the compression side, thereby allowing sensing of the flexural stress [2].

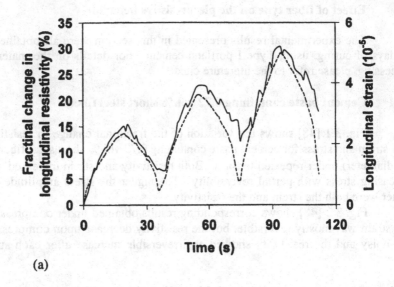

(a)

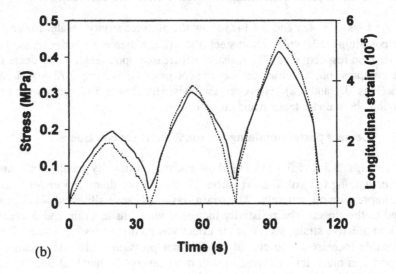

(b)

Fig. 3.1   Variation of the fractional change in electrical resistivity (solid curve) with strain (dashed curve) (a), and of the strain (solid curve) with stress (dashed curve) (b), for cement paste containing 0.72 vol.% steel fibers under tension.

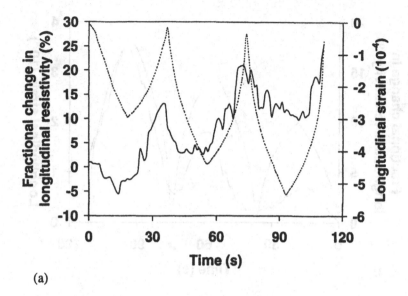

(a)

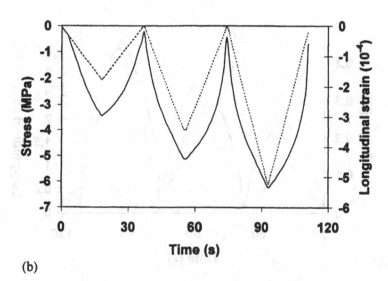

(b)

Fig. 3.2   Variation of the fractional change in electrical resistivity (solid curve) with strain (dashed curve) (a), and of the strain (solid curve) with stress (dashed curve) (b), for cement paste containing 0.72 vol.% steel fibers under compression.

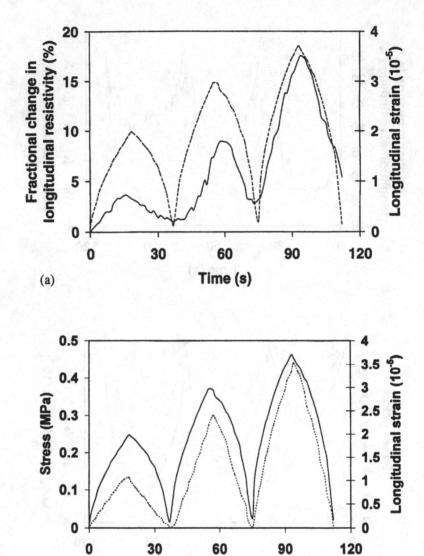

Fig. 3.3   Variation of the fractional change in electrical resistivity (solid curve) with strain (dashed curve) (a), and of the strain (solid curve) with stress (dashed curve) (b), for cement paste containing 0.36 vol.% steel fibers under tension.

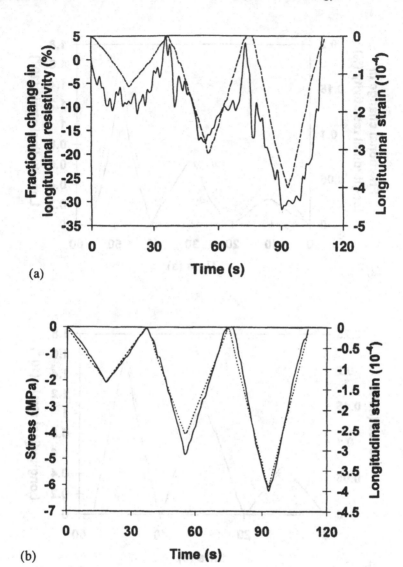

(a)

(b)

Fig. 3.4 Variation of the fractional change in electrical resistivity (solid curve) with strain (dashed curve) (a), and of the strain (solid curve) with stress (dashed curve) (b), for cement paste containing 0.36 vol.% steel fibers under compression.

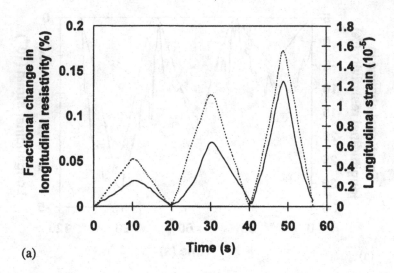

(a)

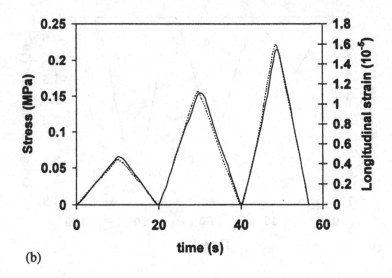

(b)

Fig. 3.5   Variation of the fractional change in electrical resistivity
                    (solid curve) with strain (dashed curve) (a), and of the
                    strain (solid curve) with stress (dashed curve) (b), for
                    cement paste containing 0.5 vol.% carbon fibers under
                    tension.

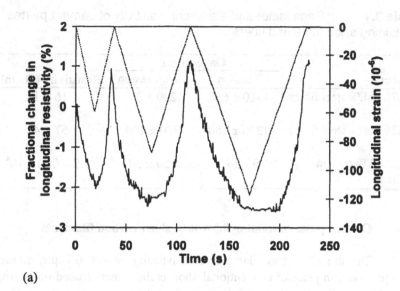

(a)

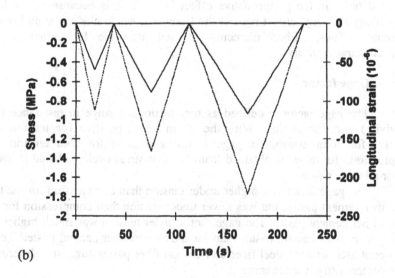

(b)

Fig. 3.6   Variation of the fractional change in electrical resistivity (solid curve) with strain (dashed curve) (a), and of the strain (solid curve) with stress (dashed curve) (b) for cement paste containing 0.5 vol.% carbon fibers under compression.

Table 3.1   Gage factor and electrical resistivity of cement pastes containing silica fume and fibers

Fibers	Gage factor		Resistivity (Ω.cm)
	Tension	Compression	
0.72 vol.% steel fibers	4560 ± 640	200 ± 30	16 ± 1
0.36 vol.% steel fibers	1290 ± 160	720 ± 100	57 ± 4
0.5 vol.% carbon fibers	90 ± 10	350 ± 30	$(1.5 \pm 0.1) \times 10^4$

3.2.4   Cement paste containing 0.5 vol.% short carbon filaments

The use of carbon filaments (catalytically grown, 0.1 µm diameter >100 µm long) in place of conventional short carbon fibers (based on isotropic pitch, 15 µm diameter, Sec. 3.2.3) in a cement-matrix composite results in increased noise in the piezoresistive effect [49]. This is because of the bent morphology and large aspect ratio of the filaments, which hinder the pull-out of filaments. Thus, carbon filaments are not attractive for cement-matrix composite strain sensors.

3.2.5   Gage factor

The gage factor is defined as the fractional change in resistance (not resistivity) per unit strain. With the strain being positive for tension and negative for compression, the gage factor is positive for both tension and compression. Its value, as obtained from the first stress cycle, is listed in Table 3.1 for all three pastes.

The gage factor was higher under tension than compression for the two steel fiber cement pastes, but was lower under tension than compression for the carbon fiber cement paste. The gage factor under tension was much higher for the two steel fiber cement pastes than for the carbon fiber cement paste. These sharp contrasts between steel fiber and carbon fiber pastes suggest a difference in the piezoresistivity mechanism.

The gage factor is higher for the steel fiber cement pastes than the carbon fiber cement paste, except for the case of the paste with 0.72 vol.% steel fibers under compression. Between the two steel fiber pastes, the value for tension is higher and that for compression is lower for the paste with a higher fiber content.

3.2.6   Electrical resistivity

The electrical resistivity of the three cement pastes of Sec. 3.2 is listed in Table 3.1. The two steel fiber pastes are much more conductive than the carbon fiber paste. This difference is because the steel fiber volume fractions are above the percolation threshold previously determined for the steel fiber case (between 0.27 and 0.36 vol.%) [46], whereas the carbon fiber volume fraction is below the percolation threshold previously determined for the carbon fiber case (between 0.5 and 1.0 vol.%) [50].

3.2.7   Discussion

Percolation means the touching of adjacent fibers so that a continuous conducting path exists. Above the percolation threshold (i.e., when percolation occurs prior to straining), the conductivity is governed by the contact resistance at the fiber-fiber contact, which is affected by tension much more than compression. Below the percolation threshold (i.e., when percolation does not occur prior to straining), the conductivity is governed by the contact resistance at the fiber-matrix interface, in case that the matrix is not insulating, i.e., the case of the cement matrix [4]. This interface is inherently weak and is thus affected by compression more than tension. Thus the piezoresistivity in steel fiber cement pastes is dominated by the effect of strain (particularly tensile strain) on the fiber-fiber contact, whereas that in carbon fiber cement is dominated by the effect of strain (particularly compressive strain) on the fiber-matrix contact.

Steel fibers are much more ductile than carbon fibers. The ductility of the steel fibers is favorable for the change in fiber-fiber contact, which involves more movement than the change in fiber-matrix contact.

An increase in the steel fiber volume fraction causes the gage factor under tension to increase, but causes that under compression to decrease. This supports the fact that, in the presence of percolation, tension has more effect on the fiber-fiber contact than compression.

Although the gage factor is relatively low for the carbon fiber cement paste than for the steel fiber pastes, the signal-to-noise ratio is higher and the reversibility upon unloading is better for the former, as shown by comparing Figs. 3.1(a) – 3.6(a). In particular, the signal-to-noise ratio is very low for the steel fiber cement pastes under compression. Therefore, the carbon fiber cement paste is superior as a strain sensor than the steel fiber counterparts. Between the two steel fiber pastes, the one with the lower fiber volume fraction (0.36%) is superior, due to better reversibility upon unloading, the higher gage factor under compression, and the better balance in gage factor between tension and compression.

The relatively higher signal-to-noise ratio and superior reversibility (upon unloading) of the carbon fiber cement paste is attributed to the relatively small movement of the fibers associated with changing the tightness of the fiber-matrix interface, compared to the relatively large movement of the fibers associated with changing the proximity between adjacent fibers.

It was previously believed that the inferior piezoresistive performance of steel fiber cement compared to carbon fiber cement was due to the large diameter (60 μm) of the steel fiber used in the previous work [36]. However, the steel fiber diameter (8 μm) was even less than the carbon fiber diameter (15 μm) in this work. Thus, the inferior performance of steel fiber cement is related to the difference in piezoresistive mechanism, rather than the difference in diameter.

3.3   Effect of curing age on the piezoresistive behavior

The piezoresistive behavior of carbon fiber-reinforced mortar changes at a curing age between 7 and 14 days. At 14 days and beyond, the electrical

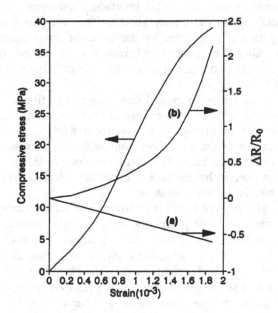

Fig. 3.7   Fractional change in resistance ($\Delta R/R_0$) versus the compressive strain during static compression of carbon fiber-reinforced mortar up to failure. (a) 28 days of curing. (b) 7 days of curing.

resistance decreases upon compression, as shown in Fig. 3.6. However, at a curing age of 7 days, it increases upon compression. The contrast is shown in Fig. 3.7, which shows the piezoresistive behavior upon compression up to failure [8]. The contrast is attributed to the effect of the curing age on the fiber-cement bond strength, which diminishes with increasing curing age from 7 to 14 days, as shown for 60 μm-diameter stainless steel fiber [51]. At 7 days, the strong bond causes the need to weaken the bond prior to fiber pull-out. At 14 days and beyond, the bond is weak to start with, so bond weakening is not necessary prior to fiber pull-out.

3.4 Mechanism behind piezoresistivity in short fiber cement-based materials

The piezoresistivity in carbon fiber cement-based materials involves a mechanism in which the fibers (discontinuous and electrically conductive) are pulled out irreversibly from the cement (less conductive) matrix [4]. The fiber pull-out is activated by straining and accompanies crack opening. The reverse, fiber push-in, accompanies crack closing. As the amount of fiber pull-out (< 1 μm) is negligible compared to the fiber length (5 mm), the fiber-matrix interface area is essentially not affected by the fiber pull-out, but the fiber-matrix contact resistivity is increased upon fiber pull-out, thus causing the overall resistivity of the composite to increase. The reversibility of the fiber pull-out is associated with the reversibility of the crack opening. This reversibility is made possible by the fact that the fiber bridges the crack. The crack volume increase alone just cannot explain the large increase in electrical resistance.

In order for a short fiber composite to have strain sensing ability using the abovementioned mechanism, the fibers must be more conducting than the matrix, of diameter smaller than the crack length and be well dispersed. Their orientations can be random, and they do not need to touch one another (i.e., percolation is not needed). Percolation refers to the situation in which the fibers touch one another, thus allowing electrical conduction to occur from one fiber directly to another fiber.

The evidence that supports the abovementioned sensing mechanism includes the following [2,4].

1. The sensing ability was present when the fibers were conducting (i.e., carbon or steel) and absent when the fibers were non-conducting (i.e., polyethylene).
2. The sensing ability was absent when fibers were absent.
3. The sensing ability occurred at low carbon fiber volume fractions which are associated with little effect of the fiber addition on the concrete's volume electrical resistivity.

4.   There was no maximum volume electrical resistivity required in order for the sensing ability to be present.

5.   The sensing ability was present when the carbon fiber volume fraction was as low as 0.2% – way below the percolation threshold, which was 1 vol.% or above, depending on the ingredients (e.g., silica fume vs. latex) used to help disperse the fibers.

6.   Fracture surface examination showed that the fibers were separate from one another.

7.   The fractional increase in electrical resistance ($\Delta R/R_o$) upon straining did not increase with increasing carbon fiber volume fraction, even though the increase in fiber volume fraction beyond the percolation threshold caused a large decrease (by orders of magnitude) in the volume electrical resistivity.

8.   The electrical resistance increased upon straining, whether in tension or compression. In contrast, if the mechanism involved the change in proximity between adjacent fibers upon straining, the resistance would have increased in tension and decreased in compression at all curing ages.

9.   The presence of carbon fibers caused the crack height to decrease by orders of magnitude. For example, the irreversible crack height observed after deformation to 70% of the compressive strength was decreased from 100 to 1 μm by the addition of carbon fibers in the amount of 0.37 vol.%, even though the compressive strength was essentially not affected by the fiber addition.

10.   The presence of carbon fibers caused the flexural toughness and tensile ductility of the composite to increase greatly.

Points 3, 4, 7 and 8 are against the change in proximity between adjacent fibers upon straining as the mechanism. Points 9 and 10, together with prior knowledge on fiber reinforced concrete [54,55], suggest the occurrence of fiber bridging. Points 1, 2, 5 and 6 suggest that the electrical contact resistance between fiber and matrix plays an important role and that between fiber and fiber does not. All the pieces of evidence together support the abovementioned mechanism. However, further work is needed to completely prove this mechanism.

3.5   Piezoresistivity in continuous fiber cement-based materials

Continuous fibers are far more effective than short fibers for reinforcement, so advanced structural composites all use continuous fibers rather than short fibers, in spite of the high cost of continuous fibers compared to short fibers. Advanced structural composites are predominantly polymer-matrix composites, due to the low density and adhesive ability of polymers. The polymer-matrix composites are widely used for lightweight structures, such as aircraft and sporting goods. Less commonly, they are used for the repair and strengthening of concrete structures [54-61]. However, polymers are much more expensive than cement, and the adhesion of polymers to concrete and the long-term durability of polymers inside concrete are of concern. Although numerous studies have been made on the use of short fibers in concrete [62], little work has been reported on the use of continuous fibers [63-67]. In contrast to short fibers, continuous fibers cannot be incorporated in a cement mix. They need to be placed and made straight and parallel prior to the pouring of cement paste around it [68]. Thus, the preparation of continuous fiber cement-matrix composites is much more complicated than that of short fiber cement-matrix composites.

Unidirectional continuous carbon fiber reinforcement results in cement-matrix composites that exhibit tensile strength approaching that expected by calculation based on the Rule of Mixtures [64]. Due to the electrical conductivity of carbon fibers and the slight conductivity of the cement matrix, measurement of the DC electrical resistance of the composite provides a way to detect damage [68]. Fiber breakage obviously causes the longitudinal resistance to increase irreversibly. Fiber-matrix bond degradation obviously increases the transverse resistance, but it also increases the longitudinal resistance when the electrical current contacts are on the surface (e.g., perimetrically around the composite in a plane perpendicular to the longitudinal direction). When the transverse resistivity is increased, the electrical current has more difficulty in penetrating the entire cross section of the specimen, thereby resulting in an increase in the measured longitudinal resistance. Note that the electrical resistivity of carbon fibers is 10^{-4} Ω.cm, whereas that of cement paste is 10^5 Ω.cm.

Fig. 3.8 [68] shows the relationship between stress and strain and that between fractional resistance change ($\Delta R/R_o$) and strain during static tensile testing up to failure for a composite with 2.57 vol. % carbon fibers (continuous, 11 µm diameter). The stress-strain curve is linear up to a strain of 0.2%, at which the resistance starts to increase abruptly. Fig. 3.9 [68] shows the variation of $\Delta R/R_o$ during loading and unloading for various stress amplitudes within the linear portion of the stress-strain curve for a specimen with essentially the same fiber content. The resistance increases upon loading and decreases upon unloading in every cycle, such that the resistance increase is not totally reversible. The gage factor, which is the fractional change in resistance (reversible portion) per unit strain, is 28, 21 and 17 for the first, second and third cycles respectively (Fig. 3.9). The decrease in gage factor with increasing cycle number (increasing stress amplitude) (Table 3.2) is attributed to the decrease in reversibility with increasing stress amplitude. It is not clear why the intermediate fiber volume fraction gives the highest gage factor. Investigation of composites with different fiber contents shows that the extent of

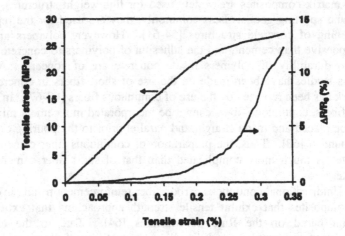

Fig. 3.8   Relationship between stress and strain and that between fractional resistance change ($\Delta R/R_o$) and strain during static tensile testing up to failure for a cement-matrix composite with 2.57 vol. % continuous carbon fibers.

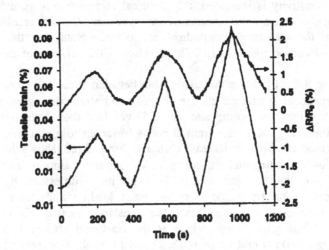

Fig. 3.9   Variation of $\Delta R/R_o$ during loading and unloading for various stress amplitudes within the linear portion of the stress-strain curve for a cement-matrix composite with 2.60 vol. % continuous carbon fibers.

Table 3.2   Gage factor

Cycle No.	Maximum load (lb)	Fiber volume fraction (%)		
		2.60 ± 0.06	5.14 ± 0.25	7.24 ± 0.24
1	50	32.6 ± 7.9	57.6 ± 0.06	33.7 ± 6.5
2	100	24.6 ± 6.9	41.7 ± 2.6	24.0 ± 2.0
3	150	16.3 ± 1.3	40.9 ± 1.7	23.4 ± 3.6

Table 3.3   Tensile properties and electrical resistivity

	Carbon fiber volume fraction (%)		
	2.57 ± 0.42	5.19 ± 1.35	7.37 ± 1.17
Tensile strength (MPa)			
Measured	27.2 ± 1.2	57.3 ± 1.1	85.7 ± 1.32
Calculted*	30.8	64.4	98
Tensile modulus (GPa)			
Measured	11.1 ± 0.52	14.6 ± 0.86	17.3 ± 0.92
Calculted*	13.1	17.1	20.8
Ductility (%)	0.341 ± 0.011	0.468 ± 0.008	0.485 ± 0.008
Resistivity (Ω.cm)			
Measured	$(1.10 \pm 0.11) \times 10^{-1}$	$(8.40 \pm 0.94) \times 10^{-2}$	$(4.56 \pm 1.32) \times 10^{-2}$
Calculted*	5.91×10^{-2}	2.83×10^{-2}	1.86×10^{-2}

* Based on the Rule of Mixtures.

irreversibility in resistance increase is greater when the stress amplitude as a fraction of the tensile strength is higher.

Similar piezoresistive behavior was observed for composites with various fiber contents [68]. Table 3.3 lists the tensile properties and resistivity of composites with various fiber contents. The tensile strength and modulus approach the values calculated based on the Rule of Mixtures. The resistivity is higher than that calculated from the Rule of Mixtures. The ductility, strength and modulus all increase with increasing fiber volume fraction.

The abrupt increase in resistance at high strains is accompanied by a decrease in modulus (Fig. 3.8), so it is attributed to fiber breakage. The smaller increase in resistance at low strains is not accompanied by any change in modulus (Fig. 3.8), so it is attributed to fiber-matrix interface degradation. The degradation causes the fiber-matrix contact resistivity to increase, thereby affecting the measured resistance, as explained above. Fig. 3.9 shows that the resistance increase due to fiber-matrix interface degradation is mostly reversible. The large gage factor means that the resistance increase cannot be explained by the dimensional change, which would have resulted in a gage factor of 2 only.

The partly reversible fiber-matrix interface degradation probably involves reversible slight loosening of the interface. The irreversible part of the resistance increase is associated with irreversible degradation of the interface. The reversibility is consistent with that observed in short carbon fiber cement-matrix composites. The reversible resistance change means that the continuous carbon fiber composites are strain sensors. The mechanism of reversible resistance increase is fiber-matrix interface loosening for both short fiber and continuous fiber composites. However, the gage factor is much higher for short fiber (Table 3.1) than continuous fiber composites.

In spite of the effort to align the fibers, the fiber alignment is not perfect, as shown by the low strength, low modulus and high resistivity relative to the calculated values (Table 3.3). Nevertheless, the tensile strength, which reaches 86 MPa, makes these composites attractive for structural applications related to tension members, repair, surface strengthening and lightweight structures.

Piezoresistivity also occurs in continuous carbon fiber epoxy-matrix composites [69]. However, the resistance of the epoxy-matrix composites in the fiber direction decreases upon tension in the fiber direction, whereas that of the cement-matrix composites increases upon tension in the fiber direction. This difference in behavior is due to the difference in mechanism. The resistance decrease in the epoxy-matrix composites is due to the increase in the degree of fiber alignment [69], whereas that in the cement-matrix composites is due to the fiber-matrix interface degradation. The fiber-matrix bond is much stronger for epoxy than cement, and the fiber content is much higher for epoxy- than cement-matrix composites. Moreover, epoxy is much more ductile than cement under tension. These differences in characteristics between epoxy and cement probably cause the difference in piezoresistive behavior.

3.6   Piezoresistivity in short fiber cement coating

Because most structures are not built with carbon fiber-reinforced concrete but with conventional concrete, the applicability of a cement-based material as a strain sensor can be widened by using the material as a strain sensing coating on conventional concrete. Cement paste containing short carbon fibers is an effective strain-sensing coating, as tested when the coating (with fibers, 5 mm thick) is on either the tension side or the compression side of a cement specimen (without fiber) under flexure [70]. The resistance is measured with surface electrical contacts on either side (four-probe method with four electrical contacts on the same surface). The resistance increases reversibly on the tension side upon loading and decreases reversibly on the compression side upon loading, such that the magnitude of the fractional change in resistance is higher and the resistance change is less noisy at the tension side than at the compression side. This means that the tension side (i.e., the bottom side of a slab loaded on the top surface) is preferred to the compression side for sensing in practical structures. Fig. 3.10 shows the fractional change in resistance during cyclic flexure for the case of the strain sensing coating being carbon fiber latex cement paste on the tension side. The resistance is irreversibly increased after the first cycle, probably due to minor damage. The behavior is similar whether the strain sensing coating contains silica fume or latex.

3.7   DC vs. AC

The data presented in all the above sections of this chapter involve the use of DC electrical power, i.e., the DC electrical resistance is measured. Under AC conditions, the impedance Z consists of the resistance R_s (real part of Z) and the reactance X_s (imaginary part of Z), i.e., $Z = R_s + iX_s$, where the subscript s refers to a configuration in which the sample is in series connection with the measuring circuit. AC provides both resistance and reactance information, and AC is relevant to data acquisition by wireless methods. It has been found that in carbon fiber (short)-reinforced mortar at 7 days of curing, the reactance X_s is a more sensitive indicator than the resistance R_s, as the fractional change in reactance exceeds the fractional change in resistance upon deformation [7]. The effect of strain on the reactance relates to the effect of strain on the polarization [71].

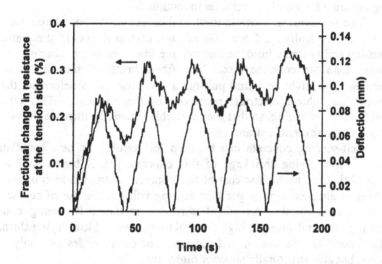

Fig. 3.10   Fractional change in resistance and deflection during cyclic flexure for the case of the strain sensing coating being carbon fiber latex cement paste on the tension side.

3.8   Application in weighing in motion

Traffic monitoring, an essential part of traffic control and management, involves real-time monitoring and requires strain sensors, which may be optical, electrical, magnetic or acoustic. The sensors are conventionally attached to or embedded in the highway for which traffic monitoring is desired. The sensors suffer from (i) their sensing ability being limited to their immediate vicinities, (ii) they are not sufficiently durable, and (iii) they are too expensive for widespread use. A relatively new technology involves the use of concrete itself as the sensor, so that no embedded or attached sensor is needed. Because the structural material is also a sensor, the whole structure is sensed and the sensor (just concrete) is durable and inexpensive. Hence, all three problems described above for conventional sensors are removed by the use of this self-sensing concrete.

The weighing of vehicles such as trucks is necessary to avoid damage to highways due to overweight vehicles. It is currently conducted in weighing stations off the highway while the vehicle is stationary. The monitoring of the weight of vehicles can be more convenient and effective if the weighing is done on the highway while the vehicle is moving normally. In this way, traffic is not affected and time is saved. If the whole highway is capable of weighing, the monitoring is continuous and hence is more thorough than the current method. This section describes a laboratory demonstration of the effectiveness of self-sensing concrete for weighing vehicles in motion.

The self-sensing concrete used in this section is concrete containing a small amount (typically 0.2-0.5 vol.%) of short carbon fibers (15 μm diameter). The sensing ability stems from the fact that the fibers are much more electrically conducive than the concrete mix. The fibers bridging microcracks in the concrete undergo slight (< 1 μm) pull-out as the concrete is deformed, thereby increasing the volume electrical resistivity of the concrete. This effect is reversible upon unloading, so that the reversible resistivity increase provides an indication of the reversible strain.

Self-sensing concrete can be used for retrofits or new installations. Due to the low drying shrinkage of this concrete [2], it bonds well to old concrete [72]. Due to the low cost of this concrete compared to concrete with embedded or attached sensors per unit sensing volume, the use of self-sensing concrete is economical. An added attraction is that self-sensing concrete exhibits high flexural strength, high flexural toughness and low drying shrinkage [2,73]. Therefore, the use of self-sensing concrete provides not only smart highways, but also structurally superior highways.

In a laboratory demonstration of the use of the self-sensing concrete (with fine and coarse aggregates) for weighing in motion [74], a vertical wheel (a car tire) is allowed to rotate against the cylindrical surface of two horizontal concrete rollers (corresponding to the highway), as illustrated in Fig. 3.11. The

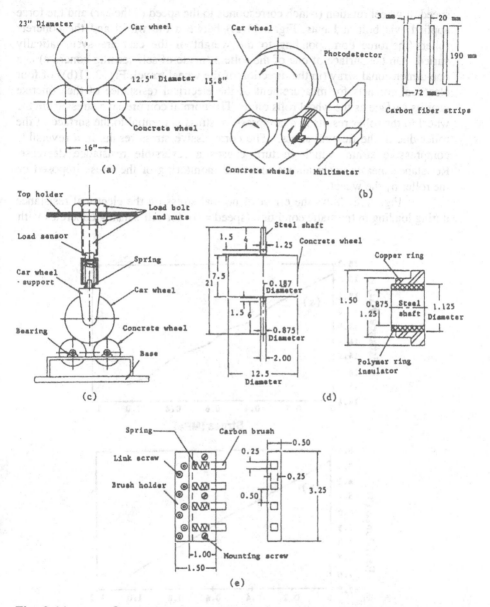

Fig. 3.11   Set-up for testing weighing in motion. Dimensions are in inches unless stated otherwise. (a) Two-dimensional view. (b) Three-dimensional view and electrical contact geometry. (c) Loading mechanism. (d) Steel shaft and copper slip ring. (e) Carbon brush assembly.

speed of wheel rotation (which corresponds to the speed of the car) and the force applied (via bolt and nuts, Fig. 3.11(c)) between the wheel and the concrete rollers (the force corresponding to the weight of the car) are systematically varied. On the surface of one of the rollers (made of self-sensing concrete) is a one-dimensional array (in the direction of the wheel travel, Fig. 3.11(b)) of four electrical contacts for measurement of the electrical resistance of the concrete near its surface as the wheel rolls on it. The normal compressive stress from the wheel to the roller results in a compressive stress tangential to the surface of the roller due to the flexural stress. The compressive stress results in a reversible compressive strain, which in turn causes a reversible resistance decrease. Resistance measurement allows real-time monitoring of the stress imposed on the roller by the wheel.

Fig. 3.12 shows the effect of normal stress on the electrical resistance during loading in the static condition (speed = 0) for self-sensing concretes with

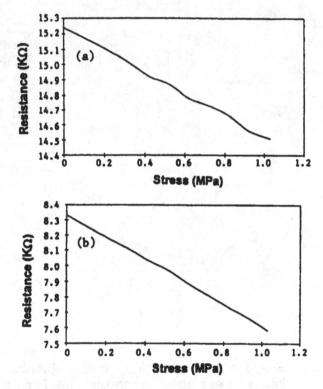

Fig. 3.12   Effects of stress on resistance at zero speed for self-sensing concrete with carbon fibers in amounts of (a) 0.5% and (b)1.0% by weight of cement.

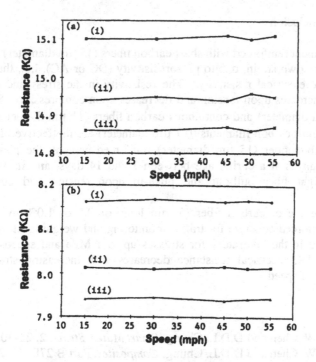

Fig. 3.13   Effect of speed on resistance at a constant stress for self-sensing concrete with carbon fibers in amounts of (a) 0.5% and (b) 1.0% by weight of cement. (i) 0.21 MPa. (ii) 0.42 MPa. (iii) 0.52 MPa.

carbon fibers in amounts of 0.5% and 1.0% by weight of cement. The resistance decreases with increasing stress and with increasing fiber content. The resistance change is totally reversible at all stress levels, as shown by the data obtained after unloading at each stress level. Fig. 3.13 shows the effect of speed for self-sensing concretes with carbon fibers in amounts of 0.5% and 1.0% by weight of cement. At each stress level, the resistance is independent of speed. The resistance change is totally reversible at all stress levels and at all speeds.

Fig. 3.12 and 3.13 mean that self-sensing concrete can monitor traffic and weigh vehicles in motion by sensing in the real time the normal stress due to vehicles up to at least 1 MPa at vehicle speeds up to at least 55 mph. The output is the electrical resistance, which decreases reversible with increasing stress and is independent of speed. The concrete contains short carbon fibers in amounts of 0.5 or 1.0% by weight of cement. The higher fiber content yields lower resistance and less noise, but it involves higher cost.

3.9   Conclusion

Cement reinforced with short carbon fibers (15 μm diameter) is capable of sensing its own strain, due to piezoresistivity (DC or AC), i.e., the effect of strain on the electrical resistivity. The resistivity in the stress and transverse directions increases upon tension and decreases upon compression. Short steel fibers (8 μm diameter) and continuous carbon fibers (11 μm diameter) are less effective. Short carbon filaments (0.1 μm diameter)-are ineffective. In the case of short carbon fiber (15 μm diameter) reinforced cement, the piezoresistive behavior changes at a curing age between 7 and 14 days, and its mechanism involves slight fiber pull-out and push-in upon tension and compression respectively.

The use of carbon fiber (5 mm long, or 0.5 or 1.0% by weight of cement)-reinforced concrete for traffic monitoring and weighing in motion was demonstrated in the laboratory for stresses up to 1 MPa and speeds up to 55 mph. The DC electrical resistance decreases with increasing stress and is independent of speed.

References

1.   P.-W. Chen and D.D.L. Chung, *Smart Mater. Struct.* 2, 22-30 (1993).
2.   P.-W. Chen and D.D.L. Chung, *Composites*, Part B 27B, 11-23 (1996).
3.   P.-W. Chen and D.D.L. Chung, *J. Amer. Ceramic Soc.* 78(3), 816-818 (1995).
4.   D.D.L. Chung, *Smart Mater. & Struct.* 4, 59-61 (1995).
5.   P.-W. Chen and D.D.L. Chung, *ACI Mater. J.* 93(4), 341-350 (1996).
6.   X. Fu and D.D.L. Chung, *Cem. Conc. Res.* 26(1), 15-20 (1996).
7.   X. Fu, E. Ma, D.D.L. Chung and W.A. Anderson, *Cem. Concr. Res.* 27(6), 845-852 (1997).
8.   X. Fu and D.D.L. Chung, *Cem. Concr. Res.* 27(9), 1313-1318 (1997).
9.   X. Fu, W. Lu and D.D.L. Chung, *Cem. Concr. Res.* 28(2), 183-187 (1998).
10.   X. Fu, W. Lu and D.D.L. Chung, *Carbon* 36(9), 1337-1345 (1998).
11.   M. Qizhao, Z. Binyuan, S. Darong and L. Zhuoqiu, *J. Wuhan Univ. of Tech.* 11(3), 41-45 (1996).
12.   Z.-Q. Shi and D.D.L. Chung, *Cem. Concr. Res.* 29(3), 435-439 (1999).
13.   S. Wen and D.D.L. Chung, *Cem. Concr. Res.* 31(2), 297-301 (2001).
14.   S. Wen and D.D.L. Chung, *Cem. Concr. Res.* 30(8), 1289-1294 (2000).
15.   P. Robins, S. Austin, J. Chandler and P. Jones, *Cem. Concr. Res.* 31(5), 719-729 (2001).
16.   C.S. Wang, F. Wu and F.-K. Chang, *Smart Mater. Struct.* 10(3), 548-552 (2001).
17.   X. Wang and D.D.L. Chung, *Smart Mater. Struct.* 5, 796-800 (1996).

18.   X. Wang and D.D.L. Chung, *Smart Mater. Struct.* 6, 504-508 (1997).

19.   X. Wang and D.D.L. Chung, *Polym. Compos.* 18(6), 692-700 (1997).

20.   X. Wang and D.D.L. Chung, *Composites: Part B* 29B(1), 63-73 (1998).

21.   P.E. Irving and C. Thiagarajan, *Smart Mater. Struct.* 7, 456 (1998).

22.   N. Muto, H. Yanagida, T. Nakatsuji, M. Sugita, Y. Ohtsuka, Y. Arai and C. Saito, *Adv. Compos. Mater.* 4(4), 297-308 (1995).

23.   M. Sugita, H. Yanagida and N. Muto, *Smart Mater. Struct.* 4(1A), A52-A57 (1995).

24.   R. Prabhakaran, *Experimental Techniques* 14(1), 16-20 (1990).

25.   K. Schulte and Ch. Baron, *Compos. Sci. Tech.* 36, 63-76 (1989).

26.   K. Schulte, *J. Physique IV, Colloque C7* 3, 1629-1636 (1993).

27.   J.C. Abry, S. Bochard, A. Chateauminois, M. Salvia and G. Giraud, *Compos. Sci. Tech.* 59(6), 925-935 (1999).

28.   A.S. Abry, S. Bochard, A. Chateauminois, M. Salvia and G. Giraud, *Compos. Sci. Tech.* 59(6), 925-935 (1999).

29.   A.S. Kaddour, F.A.R. Al-Salehi, S.T.S. Al-Hassani and M.J. Hinton, *Compos. Sci. Tech.* 51(3), 377-385 (1994).

30.   J. Kost, M. Narkis and A. Foux, *J. Appl. Polym. Sci.* 3937-3946 (1984).

31.   S. Radhakrishnan, S. Chakne and P.N. Shelke, *Mater. Lett.* 18, 358-362 (1994).

32.   P.K. Pramanik, D. Khastgir, S.K. De and T.N. Saha, *J. Mater. Sci.* 25, 3848-3853 (1990).

33.   M. Taya, W.J. Kim and K. Ono, *Mechanics of Materials* 28(1/4), 53-59 (1998).

34.   A. Ishida, M. Miyayama and H. Yanagida, *J. Am. Ceramic Soc.* 77(4), 1057-1061 (1994).

35.   J. Cao, S. Wen and D.D.L. Chung, *J. Mater. Sci.* 36(18), 4351-4360 (2001).

36.   P.-W. Chen and D.D.L. Chung, *ACI Mater. J.* 93(4), 341-350 (1996).

37.   P.-W. Chen and D.D.L. Chung, *ACI Mater. J.* 93(2), 129-133 (1996).

38.   M. Teutsch, *Betonwerk und Fertigteil-Technik* 67(4), 56-63 (2001).

39.   Z. Bayasi and H. Kaiser, *Indian Concr. J.* 75(3), 215-222 (2001).

40.   M.C. Nataraja, N. Dhang and A.P. Gupta, *Indian Concr. J.* 75(4), 287-290 (2001).

41.   A. Alavizadeh-Farhang and J. Silfwerbrand, *Transportation Research Record* (1740), 25-32 (2000).

42.   B. Lotfy, *J. Eng. & Appl. Sci.* 48(3), 455-471 (2001).

43.   W. Yao, J. Li and K. Wu, *Cem. Concr. Res.* 33(1), 27-30 (2003).

44.   N. Febrillet, A. Kido, Y. Ito and K. Ishibashi, *Transactions of the Japan Concrete Institute* 22, 243-252 (2000).

45.   W. Sun, H. Chen, X. Luo and H. Qian, *Cem. Concr. Res.* 31(4), 595-601 (2001).

46.   S. Wen and D.D.L. Chung, *Cem. Concr. Res.*, in press.

47.   S. Wen and D.D.L. Chung, *J. Electronic Mater.* 30(11), 1448-1451 (2001).
48.   S. Wen and D.D.L. Chung, *Adv. Cem. Res.*, in press.
49.   X. Fu and D.D.L. Chung, *Cem. Concr. Res.* 26(10), 1467-1472 (1997).
50.   P.-W. Chen and D.D.L. Chung, *J. Electronic Mater.* 24(1), 47-51 (1995).
51.   X. Fu and D.D.L. Chung, *Cem. Concr. Res.* 26(1), 15-20 (1996).
52.   V.C. Li, Y. Wang and S. Backer, *J. Mech. Phys. Solids* 39(5), 607-625 (1991).
53.   V.C. Li, *ASCE J. Mater. Civil Eng.* 4(1), 41-57 (1992).
54.   T. Norris, H. Saadatmanesh and M.R. Ehsani, *J. Struct. Eng.* 123(7), 903-911 (1997).
55.   A.Z. Fam, S.H. Rizkalla and G. Tadros, *ACI Struct. J.* 94(1), 77-86 (1997).
56.   H. Yoshizawa, T. Myojo, M. Okoshi, M. Mizukoshi and H.S. Kliger, *Proc. Mater. Eng. Conf. On Materials for the New Millennium*, ASCE, New York, NY, 2, 1608-1616 (1996).
57.   K. Takeda, Y. Mitsui and K. Murakami, *Composites – Part A* 27(10), 981-987 (1996).
58.   C.A. Ballinger, *Int. SAMPE Symp. Exhib. (Proc.)*, SAMPE, Covina, CA, 42(2), 927-932 (1997).
59.   A.A. Abdelrahman and S.H. Rizkalla, *ACI Struct. J.* 94(4), 447-457 (1997).
60.   M. Missihoun, I. M'Bazaa and P. Labossiere, *Annual Conf. – Canadian Soc. Civil Eng.*, Canadian Society for Civil Engineering, Montreal, Que., Can., 6, 181-189 (1997).
61.   K.A. Soudki, M.F. Green and F.D. Clapp, *Pci J.* 42(5), 78-87 (1997).
62.   N. Banthia, *ACI SP-142, Fiber Reinforced Concrete*, J.I. Daniel and S.P. Shah, Eds., ACI, Detroit, MI, 1994, p. 91-120.
63.   Q. Zheng and D.D.L. Chung, *Cem. Concr. Res.* 19, 25-41 (1989).
64.   K. Saito, N. Kawamura and Y. Kogo, *21st Int. SAMPE Technical Conf.* 1989, p. 796-802.
65.   H. Kolsch, *J. Compos. Construction* 2(2), 105-109 (1998).
66.   T. Uomoto, *Adv. Compos. Mater.* 4(3), 261-269 (1995).
67.   A. Pivacek, G.J. Haupt and B. Mobasher, *Adv. Cem. Based Mater.: ACBM* 6(3-4), 144-152 (1997).
68.   S. Wen, S.Wang and D.D.L. Chung, *J. Mater. Sci.* 35(14), 3669-3676 (2000).
69.   X. Wang and D.D.L. Chung, *Smart Mater. Struct.* 5, 796-800 (1996).
70.   S. Wen and D.D.L. Chung, *Cem. Concr. Res.* 31(4), 665-667 (2001).
71.   S. Wen and D.D.L. Chung, *Cem. Concr. Res.* 31(2), 291-295 (2001).
72.   P. Chen, X. Fu and D.D.L. Chung, *Cem. Concr. Res.* 25(3), 491-496 (1995).

73.   P. Chen and D.D.L. Chung, *Composites* 24(1), 33-52 (1993).
74.   Zeng-Qiang Shi and D.D.L. Chung, *Cem. Concr. Res.* 29(3), 435-439 (1999).

4

Cement-Based Materials for Damage Sensing

4.1   Damage sensing

Damage sensing (i.e., structural health monitoring) is valuable for structures for the purpose of hazard mitigation. It can be conducted during the damage by acoustic emission detection. It can also be conducted after the damage by ultrasonic inspection, liquid penetrant inspection, dynamic mechanical testing or other techniques. Real-time monitoring gives information on the time, load condition or other conditions at which damage occurs, thereby facilitating the evaluation of the cause of the damage. Moreover, real-time monitoring provides information as soon as damage occurs, thus enabling timely repair or other hazard precaution measures.

Real-time monitoring allows study of the damage evolution, which refers to how damage evolves in a damaging process and is the subject of much modeling work [1-12], due to its fundamental importance in relation to the science of damage. Limited experimental observation of the damage evolution has involved the use of acoustic emission [13-15], thermoelastic stress analysis [16] and computer tomography (CT) scanning [17]. In contrast, this chapter uses electrical resistance measurement, which is advantageous in its sensitivity to even minor, microscopic and reversible effects.

Stress application can generate defects, which may be a form of damage in a material. Stress application can also heal defects, particularly in the case of the stress being compressive. This healing is induced by stress [18] and is to be distinguished from healing that is induced by liquids, chemicals or particles [19-29]. On the other hand, stress removal can aggravate defects, particularly in the case of the stress being compressive and the material being brittle. The generation, healing and aggravation of defects during dynamic loading are referred to as defect dynamics. The little prior attention on defect dynamics is mainly due to the dynamic nature of defect healing and aggravation. For example, stress application can cause healing, and subsequent unloading can cancel the healing. This reversible nature of the healing makes the healing observable only in real time during loading. On the other hand, defect generation tends to be irreversible upon unloading, so it does not require observation in real time.

Observation in real time during loading is difficult for microscopy, particularly transmission electron microscopy, which is the type of microscopy that is most suitable for the observation of microscopic defects. However,

149

observation in real time during loading can be conveniently performed by electrical measurement. As defects usually increase the electrical resistivity of a material, defect generation tends to increase the resistivity whereas defect healing tends to decrease the resistivity.

The strain rate affects the damage evolution during static stress application, in addition to affecting the mechanical properties in the case of a viscoelastic material. Real-time monitoring by electrical resistivity measurement during straining at various rates allows study of the effect of strain rate on the damage evolution. Similarly, real-time electrical resistivity measurement allows monitoring of the damage evolution during creep, drying shrinkage, fatigue and freeze-thaw cycling.

Damage in cement-based materials is most commonly studied by destructive mechanical testing after different amounts of damage. However, this method does not allow the monitoring of the progress of damage on the same specimen and is not sufficiently sensitive to minor damage. As different specimens can differ in the flaws, damage evolution is more effectively studied by monitoring one specimen throughout the process rather than interrupting the process at different times for different specimens. However, the monitoring of one specimen throughout the process requires a nondestructive method that is sensitive to minor damage. Electrical resistivity measurement is effective for damage monitoring, particularly in the regime of minor damage, in addition to monitoring both defect generation and defect healing in real time.

Damage sensing should be distinguished from strain sensing (Ch. 3), as strain can be reversible and is not necessarily accompanied by damage. Damage can occur within a cement-based material, at the interface between concrete and steel rebar, at the interface between old concrete and new concrete (as encountered in the use of new concrete for repair of an old concrete structure), at the interface between unbonded concrete elements and at the interface between concrete and its carbon fiber epoxy-matrix composite retrofit. This chapter addresses all these aspects of damage and is focused on the use of electrical resistance measurement for sensing damage, whether the damage is due to edge proximity, static stress, dynamic stress, freeze-thaw cycling, creep or drying shrinkage.

4.2   Sensing damage due to edge proximity

Although much work has been conducted by numerous workers for decades concerning the properties of concrete or cement mortar before and after loading, much less attention has been given to the spatial distribution of the properties, i.e., how a property varies within a single piece of concrete or mortar. For the purpose of detecting damage in a concrete structure after loading or aging, the spatial distribution of the elastic modulus and of the electromagnetic, acoustic and thermal behavior has been previously measured in a macroscopic scale (e.g., meter scale) [30-39]. However, little attention has been given to the spatial distribution of a property prior to loading, especially in a less macroscopic scale (e.g., cm or mm scale) that reflects the material behavior. In this section, the spatial distribution of the mechanical and electrical properties of mortar prior to loading is addressed. Information on the spatial distribution is fundamental to the study of concrete properties and is technologically useful to the design of concrete structural components.

The DC volume electrical resistivity distribution was determined by using the four-probe method (unless noted otherwise), using silver paint in conjunction with copper wires as electrical contacts and using a multimeter [40]. In this method, the outer two probes were for passing the current, while the inner two probes were for voltage measurement.

In order to measure the electrical resistivity and determine its spatial variation in the same direction, nine electrical contacts were made on a mortar cube, such that each contact was around the whole perimeter of the 51 x 51 x 51 mm (2 x 2 x 2 in) cube in the plane perpendicular to the direction of resistivity measurement, as illustrated in Fig. 4.1 [40]. The outset two contacts were for passing current. All the remaining contacts were successively used in pairs as voltage contacts. For example, by using the second and third contacts as voltage contacts, the resistance between these two contacts was measured; by using the third and fourth contacts as voltage contacts, the resistance between these two contacts was measured. The nine contacts allowed the resistivity to be measured at six points along the direction of resistivity measurement.

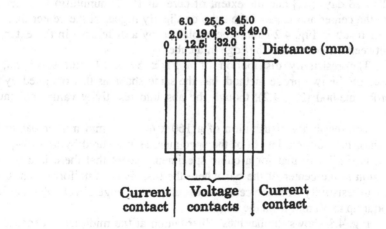

Fig. 4.1   Specimen configuration for measuring the resistivity and its distribution in the same direction. All distances are in mm.

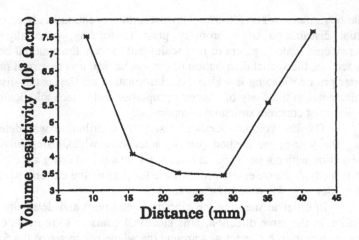

Fig. 4.2   Resistivity distribution in the horizontal direction of a 51 x 51 x 51 mm mortar cube, measured by the four-probe method.

Fig. 4.2 [40] shows the resistivity distribution. The resistivity is lower at the center than at the edges. The difference in resistivity between center and edge is large – the resistivity at the edge is about double that at the center. As the resistivity of mortar increases by only 63% when the curing time increases from 1 to 28 days [41] and the extent of cure (at 100% humidity) is either the same at the center and edge of the cube or slightly higher at the center than the edge, the result in Fig. 4.2 cannot be explained by a difference in the extent of cure between the center and the edge of a cube.

The resistivity distribution of a 51 x 51 x 51 mm specimen, as measured by the two-probe method, has the same shape as that obtained by the four-probe method (Fig. 4.2), though the absolute resistivity values are much higher.

The resistivity distribution of a 160 x 40 x 40 mm mortar bar in the direction of the 160-mm length of the specimen, as measured by the two-probe method, is similar to that for a cubic specimen, except that there is a 60-mm long region at the center of the bar where the resistivity is uniform. Thus, the increase in resistivity from the center to the edge is an edge effect, which affects the mortar up to 50 mm from the edge.

Fig. 4.3 shows the hardness distribution at the mid-plane (exposed by cutting the cube into two halves) of a 51 x 51 x 51 mm mortar cube. The hardness is higher at the center than the edge of the cube.

The increase in resistivity from center to edge is too large to be accounted for by the residual stress resulting from a difference in drying

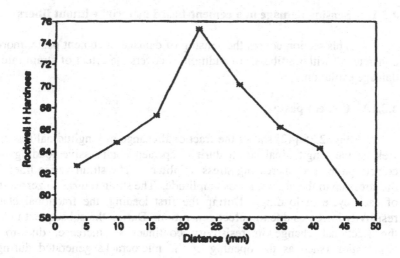

Fig. 4.3   Hardness distribution in a mortar cube.

shrinkage rate between the center and the edge. It is also too large to be accounted for by a difference in void or microcrack content between the center and edge. This effect is attributed to the faster drying at the edge than the center and the resulting poorer bond between cement paste and sand at the edge than the center. A poorer bond is associated with a higher contact electrical resistivity. This explanation is supported by the observation that the hardness is lower at the edge than the center. The depth of penetration of this edge effect is ~5 cm for the case of curing at 100% relative humidity. At a lower humidity, the difference in property between edge and center is expected to be larger than those mentioned here.

4.3 Sensing damage due to stress in a cement-based material

This section covers the sensing of damage due to stress in a cement-based material that contains no fiber admixture and in one that contains an electrically conductive fiber admixture. The fiber enhances the damage sensing ability. In addition, it covers the sensing of damage at the interface between concrete and steel rebar and the interface between new concrete and old concrete.

4.3.1   Sensing damage in a cement-based material without fibers

This section covers the sensing of damage in cement paste, mortar and concrete, all without fibers. In addition, it covers the effect of strain rate on the damage evolution.

4.3.1.1   Cement paste

Fig. 2.36 [42] shows the fractional change in longitudinal resistivity as well as the longitudinal strain during repeated compressive loading of plain cement paste at an increasing stress amplitude. The strain varies linearly with the stress up to the highest stress amplitude. The strain returns to zero at the end of each cycle of loading. During the first loading, the fractional change in resistivity increases due to defect generation. During the subsequent unloading, the fractional change in resistivity continues to increase, due to defect aggravation (such as the opening of the microcracks generated during prior loading). During the second loading, the resistivity decreases slightly as the stress increases up to the maximum stress of the first cycle (due to defect healing) and then increases as the stress increases beyond this value (due to additional defect generation). During unloading in the second cycle, the resistivity increases significantly (due to defect aggravation, probably the opening of the microcracks). During the third loading, the resistivity essentially does not change (or decreases very slightly) as the stress increases to the maximum stress of the third cycle (probably due to the balance between defect generation and defect healing). Subsequent unloading causes the resistivity to increase very significantly due to defect aggravation (probably the opening of the microcracks).

Fig. 4.4 [42] shows the fractional change in transverse resistivity as well as the transverse strain (positive due to the Poisson effect) during repeated compressive loading at an increasing stress amplitude. The strain varies linearly with the stress and returns to zero at the end of each cycle of loading. During the first loading and the first unloading, the resistivity increases due to defect generation and defect aggravation respectively, as also shown by the longitudinal resistivity variation (Fig. 2.36). During the second loading, the resistivity first increases (due to defect generation) and then decreases (due to defect healing). During the second unloading, the resistivity increases due to defect aggravation. During the third loading, the resistivity decreases due to defect healing. During the third unloading, the resistivity increases due to defect aggravation.

The variations of the resistivity in the longitudinal and transverse directions upon repeated loading are consistent in showing defect generation (which dominates during the first loading), defect healing (which dominates during subsequent loading) and defect aggravation (which dominates during subsequent unloading). The defect aggravation during unloading follows the defect healing during loading, indicating the reversible (not permanent) nature of the healing, which is induced by compressive stress. The defect aggravation during unloading also follows the defect generation during loading.

In spite of the Poisson effect, similar behavior was observed in the longitudinal and transverse resistivities. This means that the defects mentioned above are essentially nondirectional and that the resistivity variations are real.

Comparison of Figs. 2.36 and 4.4 shows that the increase in resistivity with strain during unloading in the second cycle is clearer and less noisy for the longitudinal resistivity than the transverse resistivity. This suggests that defect aggravation is more significantly revealed by the longitudinal resistivity than the transverse resistivity. Hence, the defects are not completely nondirectional.

Identification of the defect type has not been made. Microcracks were mentioned above just for the sake of illustration. The defects may be associated with certain heterogeneities in the cement paste.

Defects affect the mechanical properties. Therefore, mechanical testing (such as modulus measurement, which is nondestructive) can be used for studying defect dynamics. However, the modulus is not as sensitive to defect

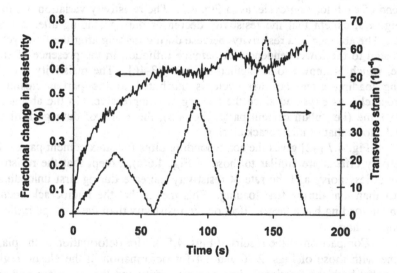

Fig. 4.4   Variation of the fractional change in transverse resistivity with time and of the transverse strain with time during dynamic compressive loading at increasing stress amplitudes within the elastic regime for plain cement paste.

dynamics as the electrical resistivity; the relationship between stress and strain is not affected while the resistivity is affected. The low sensitivity of the modulus to defect dynamics is consistent with the fact that the deformation is elastic.

Fig. 4.5(a) [42] shows the fractional change in resistivity along the stress axis as well as the strain during repeated compressive loading at an increasing stress amplitude for plain cement paste. Fig. 4.5(b) shows the corresponding variation of stress and strain during the repeated loading. The strain varies linearly with the stress up to the highest stress amplitude (Fig. 4.5(b)). The strain does not return to zero at the end of each cycle of loading, indicating plastic deformation. In contrast, Figs. 2.36 and 4.4 are concerned with effects of elastic deformation.

The resistivity increases during loading and unloading in every loading cycle (Fig. 4.5(a)). The slope of the curve of resistivity vs. time (Fig. 4.5(a)) increases with time, due to the increasing stress amplitude cycle by cycle (Fig. 4.5(b)) and the nonlinear increase in damage severity as the stress amplitude increases. The resistivity increase during loading is attributed to damage infliction. The resistivity increase during unloading is attributed to the opening of microcracks generated during loading.

Fig. 4.6 [42] gives the corresponding plots for silica fume cement paste at the same stress amplitudes as Fig. 4.5. The strain does not return to zero at the end of each loading cycle, as in Fig. 4.5. The resistivity variation is similar to Fig. 4.5, except that the resistivity decreases during loading after the first cycle. The absence of a resistivity increase during loading after the first cycle is attributed to the lower tendency for damage infliction in the presence of silica fume, which is known to strengthen cement [43-46]. The resistivity decrease during loading after the first cycle is attributed to the partial closing of microcracks, as expected since the loading is compressive. In the absence of silica fume (i.e., plain cement paste, Fig. 4.5), the effect of damage infliction overshadows that of microcrack closing.

Fig. 4.7 [42] gives the corresponding plots for latex cement paste. The resistivity effects are similar to those of Fig. 4.6(a), except that the resistivity curve is less noisy and the rate of resistivity increase during first unloading is higher than that during first loading. This means that the microcrack opening during unloading has a larger effect on the resistivity than the damage infliction during loading.

Comparison of the results of Figs. 4.5-4.7 for deformation in the plastic regime with those of Figs. 2.36 and 4.4 for deformation in the elastic regime shows that both the fractional change in resistivity and the strain are higher in the plastic regime than in the elastic regime by orders of magnitude. Another difference is that the resistivity decreases are much less significant in the plastic regime than in the elastic regime. There is no resistivity decrease at all in Fig. 4.5(a), but there are resistivity decreases in Fig. 2.36. These differences between the results of plastic and elastic regimes are consistent with the much

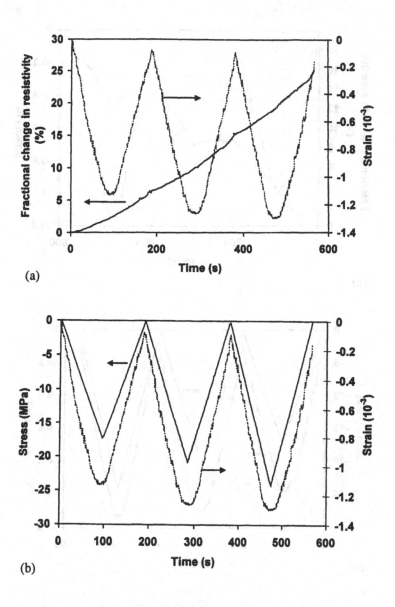

(a)

(b)

Fig. 4.5   Variation of the fractional change in electrical resistivity with time (a), of the stress with time (b), and of the strain (negative for compressive strain) with time (a,b) during dynamic compressive loading at increasing stress amplitudes for plain cement paste.

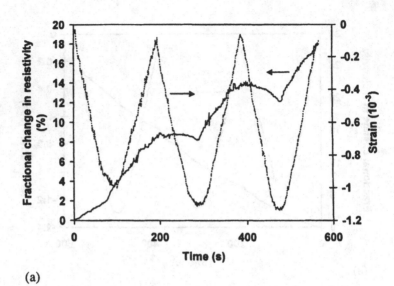

(a)

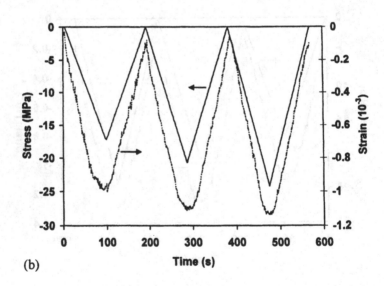

(b)

Fig. 4.6   Variation of the fractional change in electrical resistivity with time (a), of the stress with time (b), and of the strain (negative for compressive strain) with time (a,b) during dynamic compressive loading at increasing stress amplitudes for silica fume cement paste.

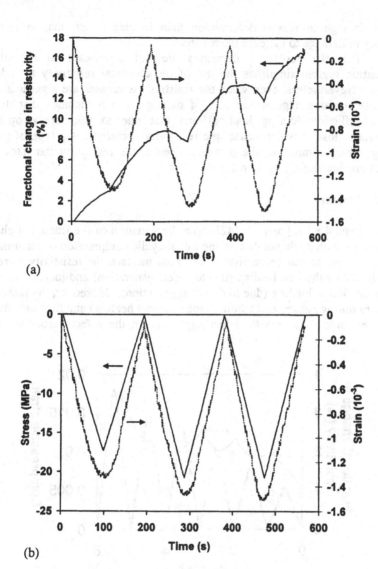

(a)

(b)

Fig. 4.7   Variation of the fractional change in electrical resistivity with time (a), of the stress with time (b), and of the strain (negative for compressive strain) with time (a,b) during dynamic compressive loading at increasing stress amplitudes for latex cement paste.

greater damage in plastic deformation than in elastic deformation and the tendency of damage to increase the resistivity.

That the resistivity decreases are not significant in the plastic deformation regime simplifies the use of the electrical resistivity to indicate damage. Nevertheless, even when the resistivity decreases are significant, the resistivity remains a good indicator of damage, which includes that due to damage infliction (during loading) and that due to microcrack opening. Microcrack closing, which causes the resistivity decreases, is a type of partial healing, which diminishes the damage. Hence, the resistivity indicates both damage and healing effects in real time.

4.3.1.2 Mortars

Figs. 4.8 [42] and 4.9 [42] show the variation of the fractional change in resistivity with cycle number during initial cyclic compression of plain mortar and silica fume mortar respectively. For both mortars, the resistivity increases abruptly during the first loading (due to defect generation) and increases further during the first unloading (due to defect aggravation). Moreover, the resistivity decreases during subsequent loading (due to defect healing) and increases during subsequent unloading (due to defect aggravation); the effect associated with

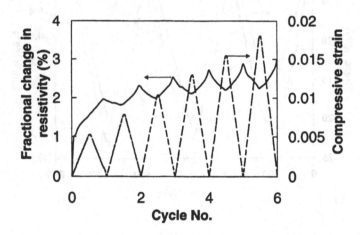

Fig. 4.8   Variation of the fractional change in resistivity with cycle no. (thick curve) and of the compressive strain with cycle no. (thin curve) during repeated compressive loading at increasing stress amplitudes within the elastic regime for plain mortar.

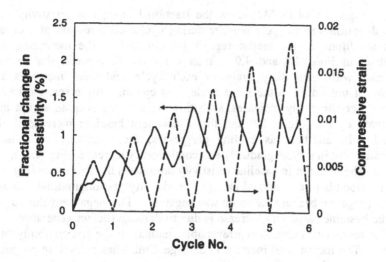

Fig. 4.9   Variation of the fractional change in resistivity with cycle no. (thick curve) and of the compressive strain with cycle no. (thin curve) during repeated compressive loading at increasing stress amplitudes within the elastic regime for silica fume mortar.

defect healing is much larger for silica fume mortar than for plain mortar. In addition, this effect intensifies as stress cycling progresses at increasing stress amplitudes for both mortars, probably due to the increase in the extent of minor damage. The increase in damage extent is also indicated by the resistivity baseline increasing gradually cycle by cycle. In spite of the increase in stress amplitude cycle by cycle, defect healing dominates over defect generation during loading in all cycles other than the first cycle.

Comparison of plain cement paste behavior (Sec. 4.3.1.1) and plain mortar behavior (this section) shows that the behavior is similar, except that the defect healing (i.e., the resistivity decrease upon loading other than the first loading) is much more significant in the mortar case. This means that the sand-cement interface in the mortar contributes significantly to the defect dynamics, particularly in relation to defect healing.

Comparison of Figs. 4.8 and 4.9 shows that silica fume contributes significantly to the defect dynamics. The associated defects are presumably at the interface between silica fume and cement, even though this interface is diffuse due to the pozzolanic nature of silica fume. The defects at this interface are smaller than those at the sand-cement interface, but this interface is large in total area due to the small size of silica fume compared to sand.

Figs. 4.10-4.12 [42] show the fractional change in resistivity in the stress direction versus cycle number during cyclic compression at a constant stress amplitude in the elastic regime (in contrast to the increasing stress amplitude in Figs. 4.8 and 4.9). Except for the first cycle, the resistivity decreases with increasing strain in each cycle and then increases upon subsequent unloading in the same cycle. As cycling progresses, the baseline resistivity continuously increases, such that the increase is quite abrupt in the first three cycles (Fig. 4.10) and that subsequent baseline increase is more gradual. In addition, as cycling progresses, the amplitude of resistivity decreasing within a cycle gradually and continuously increases (Fig. 4.10).

The increase in baseline resistivity dominates the first cycle (Fig. 4.11) and corresponds to a fractional change in resistivity per longitudinal unit strain of −1.1 (negative because the strain was negative). This negative value suggests that the baseline resistivity increase is due to damage (defect generation). The baseline resistivity increase is irreversible, indicating the irreversibility of the damage. The incremental increase in damage diminishes as cycling progresses, as shown by the baseline resistivity increasing more gradually as cycling progresses.

The reversible decrease in resistivity within a stress cycle corresponds to a fractional change in resistivity per unit strain of +0.72 at cycle number 50 (Fig. 4.12). It is attributed to defect healing (reversible) under the compressive stress. As cycling progresses, the cumulative damage (as indicated by the baseline resistivity) increases and results in a greater degree of defect healing upon compression (hence, more decrease in resistivity within a cycle).

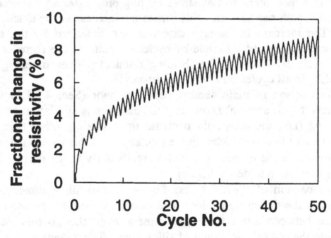

Fig. 4.10   Fractional change in resistivity and strain, both vs. compressive stress cycle number for cycles 1-3 for plain mortar.

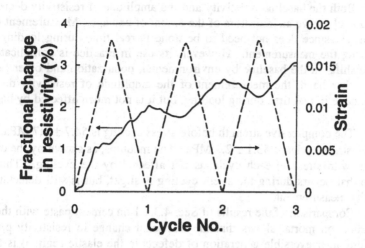

Fig. 4.11 Fractional change in resistivity and strain, both vs. compressive stress cycle number for cycles 1-3 for plain mortar.

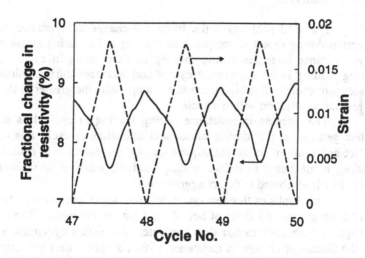

Fig. 4.12 Fractional change in resistivity and strain, both vs. compressive stress cycle number for cycles 48-50 for plain mortar.

Both the baseline resistivity and the amplitude of resistivity decreasing within a cycle serve as indicators of the extent of damage. Measurement of the baseline resistance does not need to be done in real time during loading, thus simplifying the measurement. However, its use in practice is complicated by possible shifts in the baseline by environmental, polarization and other factors. On the other hand, the measurement of the amplitude of resistivity decrease must be done in real time during loading, but it is not much affected by baseline shifts.

The compressive strength before stress cycling is 54.7 ± 1.7 MPa. That after 100 stress cycles is 53.1 ± 2.1 MPa. The modulus, as shown by the change of strain with stress in each cycle, is not affected by the cycling. Thus, the damage that occurs during the stress cycling is slight, but is still detectable by resistivity measurement.

Comparison of the results of Sec. 4.3.1.1 on cement paste with those of this section on mortar shows that the fractional change in resistivity per unit strain (due to irreversible generation of defects in the elastic regime) is higher for mortar (1.1) than for cement paste (0.10). Moreover, comparison shows that mortar is more prone to defect healing (reversible) than cement paste, as expected from the presence of the interface between fine aggregate and cement in mortar.

4.3.1.3 Concrete

Fig. 4.13 [42] shows the fractional change in resistance in the stress direction during repeated compressive loading at increasing stress amplitudes. The resistance increases during loading and unloading in cycle 1, decreases during loading in all subsequent cycles and increases during unloading in all subsequent cycles. The higher the stress amplitude, the greater is the amplitude of resistance variation within a cycle.

The increase in resistance during loading in cycle 1 is attributed to defect generation; that during subsequent unloading in cycle 1 is attributed to defect aggravation. In all subsequent cycles, the decrease in resistance during loading is attributed to defect healing, and the increase in resistance during unloading is attributed to defect aggravation.

The results of this section on concrete are consistent with those of Sec. 4.3.1.2 on mortar and those of Sec. 4.3.1.1 on cement paste. The compressive strength is higher for mortar than concrete. The defect dynamics, as indicated by the fractional change in resistance within a cycle, are more significant for concrete than mortar. The first healing, as indicated by the resistance decrease during loading in cycle 2, is much more complete for concrete than mortar. These observations mean that the interface between mortar and coarse aggregate contributes to the defect dynamics (particularly healing), due to the interfacial voids and defects.

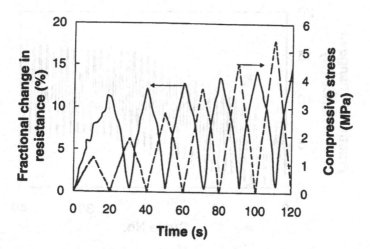

Fig. 4.13   Fractional change in resistance (solid curve) and stress (dashed curve), both vs. time during repeated compressive loading at increasing stress amplitudes, for plain concrete.

Figs. 4.14-4.16 [42] show the fractional change in resistance in the stress direction versus cycle number during cyclic compression at a constant stress amplitude. Except for the first cycle, the resistance decreases with increasing stress in each cycle and then increases upon subsequent unloading in the same cycle. As cycling progresses, the baseline resistivity gradually and irreversibly increases (Fig. 4.14). In addition, as cycling progresses, the amplitude of resistance decrease within a cycle gradually and continuously increases, especially in cycles 1-9 (Fig. 4.14).

In the first cycle, the resistance increases upon loading and unloading, in contrast to all subsequent cycles, where the resistance decreases upon loading and increases upon unloading (Fig. 4.15).

The compressive strength before stress cycling is 16.73 ± 0.86 MPa. That after 40 stress cycles is 14.24 ± 0.97 MPa. Thus, the damage that occurrs during the stress cycling is slight, but is still detectable by resistance measurement.

The gradual increase in baseline resistance as stress cycling progressed (Fig. 4.14) is attributed to irreversible and slight damage. The increase in the amplitude of resistance variation as cycling progresses (Fig. 4.14) is attributed to the effect of damage on the extent of defect dynamics. In other words, the

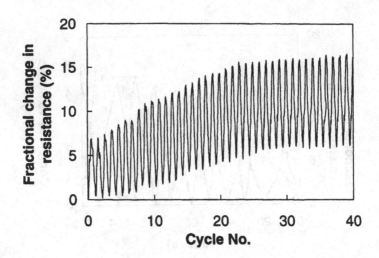

Fig. 4.14   Fractional change in resistance vs. compressive stress cycle number for cycles 1-40 for plain concrete.

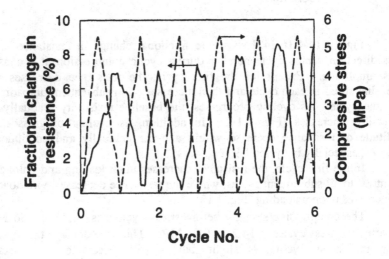

Fig. 4.15   Fractional change in resistance (solid curve) and stress (dashed curve), both vs. compressive stress cycle number for cycles 1-6 for plain concrete.

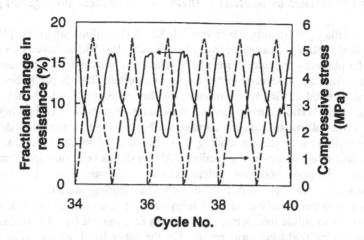

Fig. 4.16   Fractional change in resistance (solid curve) and stress (dashed curve), both vs. compressive stress cycle number for cycles 35-40 for plain concrete.

greater the damage, the greater the extent of defect healing during loading and of defect aggravation during unloading.

     The fractional loss in compressive strength after the cycling is greater for concrete than mortar, as expected from the higher compressive strength of mortar. Nevertheless, the baseline resistance increase is more significant for mortar than concrete, probably due to the relatively large area of the interface between cement and fine aggregate and the consequent greater sensitivity of the baseline resistivity to the quality of the interface between cement and fine aggregate than to the quality of the interface between mortar and coarse aggregate. In other words, the interface between cement and fine aggregate dominates the irreversible electrical effects.

4.3.1.4 Effect of strain rate

     The mechanical properties of cement-based materials are strain rate sensitive. As for most materials (whether cement-based or not), the measured strength (whether tensile or compressive) increases with increasing strain rate [47]. This effect is practically important due to the high strain rate encountered in earthquakes and in impact loading. The effect is less for high strength concrete than normal concrete [48] and is less at a curing age of 28 days than at an early age [49]. The cause of the effect is not completely understood, although it is related to the effect of strain rate on the crack propagation [47,50-52].

Although fracture mechanics [48,53,54], failure analysis [50] and mechanical testing over a wide range of strain rates [50,55,56] have been used to study the phenomenon and cause of the strain rate sensitivity of cement-based materials, the current level of understanding is limited. This is partly because of the experimental difficulty of monitoring the microstructural change during loading. Observation during loading is in contrast to that after loading. The former gives information on the damage evolution, whether the latter does not. Work on observation during loading is mainly limited to determination of the stress-strain relationship during loading. Although this relationship is important and basic, it does not give microstructural information. The use of a nondestructive real-time monitoring technique during loading is desirable. Microscopy is commonly used for microstructural observation, but it is usually not sensitive to subtle microstructural changes in a cement-based material and is not suitable for real-time monitoring. On the other hand, electrical resistivity measurement is nondestructive and fast.

Fig. 4.17 [57] shows the fractional change in resistivity in the stress direction versus the strain in the stress direction during compressive testing up to failure of cement mortar (without fiber) at three different loading rates. The

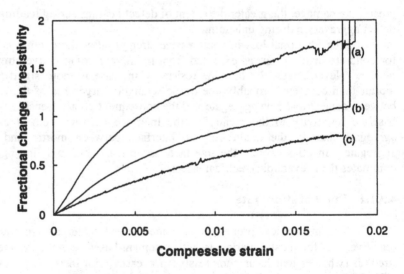

Fig. 4.17   Fractional change in resistivity vs. strain during compressive testing up to failure of mortar (without fiber) at loading rates of (a) 0.144, (b) 0.216 and (c) 0.575 MPa/s.

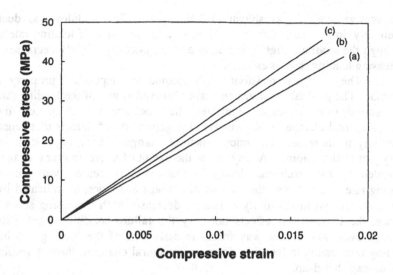

Fig. 4.18 Stress vs. strain during compressive testing up to failure of mortar (without fiber) at loading rates of (a) 0.144, (b) 0.216 and (c) 0.575 MPa/s.

Table 4.1 Effect of strain rate on the compressive properties of mortar (without fiber)

Loading rate $(MPa.s^{-1})$	Strain rate $(10^{-5} s^{-1})$	Strength (MPa)	Modulus (GPa)	Ductility (%)	Fractional change in resistivity at fracture
0.144	5.3	41.4 ± 1.6	1.83 ± 0.17	1.9 ± 0.2	1.78 ± 0.24
0.216	8.8	43.2 ± 1.0	1.85 ± 0.14	1.8 ± 0.2	1.10 ± 0.13
0.575	23.3	45.7 ± 2.1	1.93 ± 0.17	1.8 ± 0.3	0.81 ± 0.16

resistivity increases monotonically with strain and stress, such that the resistivity increase is most significant when the strain or stress is low compared to the strain or stress at fracture. Similar curvature of the resistivity curve (Fig. 4.18) occurs for all three loading rates. At fracture, the resistivity abruptly increases, as expected. Figure 4.18 [57] shows the stress vs. strain at different loading rates. The stress-strain curve is a straight line up to failure for any of the loading rates, indicating the brittleness of the failure. The higher the loading rate, the lower the fractional change in resistivity at fracture and the higher is the compressive strength, as shown in Table 4.1. The modulus and ductility essentially do not vary with the loading rate in the range of loading rate used, although the modulus slightly increases and the ductility slightly decreases with increasing loading rate, as expected.

The electrical resistivity is a geometry-independent property of a material. The gradual resistivity increase observed at any of the loading rates as the stress/strain increases indicates the occurrence of a continuous microstructural change, which involves the generation of defects that cause the resistivity to increase. The microstructural change is most significant in the early part of the loading. At any strain, the extent of microstructural change, as indicated by the fractional change in resistivity, decreases with increasing loading rate. In addition, the amount of damage at failure, as indicated by the fractional change in resistivity at failure, decreases with increasing strain rate. Hence, the loading rate affects not only the failure conditions, but also the damage evolution, all the way from the early part of the loading. A higher loading rate results in less time for microstructural changes, thereby leading to less damage build-up.

4.3.2   Sensing damage in a cement-based material containing conductive short fibers

In short fiber-reinforced concrete, the bridging of the cracks by fibers limits the crack height to values much smaller than those of concretes without fiber-reinforcement. For example, the crack height is less than 1 μm in carbon fiber reinforced mortar after compression to 70% of the compressive strength, but is about 100 μm in mortar without fibers after compression to 70% of the corresponding compressive strength (Fig. 9 of Ref. 58). As a result, the regime of minor damage is more dominant when fibers are present.

Cement reinforced with short carbon fibers is attractive due to its high flexural strength and toughness and low drying shrinkage, in addition to its strain sensing ability (Chapter 3). The strain sensing ability stems from the effect of strain on the microcrack height and the consequent slight pull-out or push-in of the fiber that bridges the crack [58]. Fiber pull-out occurs during tensile strain and causes an increase in the contact electrical resistivity at the fiber-matrix interface, thereby increasing the volume resistivity of the composite. Fiber push-in occurs during compressive strain and causes a decrease in the volume resistivity of the composite [58].

While reversible changes in electrical resistance upon dynamic loading relate to dynamic strain, irreversible changes in resistance relate to damage. The resistance of carbon fiber-reinforced cement mortar decreases irreversibly during the early stage of fatigue (the first 10% or less of the fatigue life) due to matrix damage resulting from multiple cycles of fiber pull-out and push-in [59]. The matrix damage enhances the chance of adjacent fibers to touch one another, thereby decreasing the resistivity. Beyond the early stage of fatigue and up to the end of the fatigue life, there is no irreversible resistance change, other than the abrupt resistance increase at fracture [59]. The absence of an irreversible change before fracture indicates that the mortar is not a good sensor of its fatigue damage.

Fatigue damage is to be distinguished from damage under increasing stresses. The former typically involves stress cycling at a low and fixed stress amplitude [59], whereas the latter typically involves higher stresses. The former tends to occur more gradually than the latter. Thus, the failure to sense fatigue damage [59] does not suggest failure to sense damage in general.

The sensing of damage under increasing stresses has been demonstrated in carbon fiber-reinforced concrete [60] and carbon fiber-reinforced mortar [61]. The damage is accompanied by a partially reversible increase in the electrical resistivity of the concrete. The greater the damage, the larger the resistivity increase. As fiber breakage would have resulted in an irreversible resistivity increase, the damage is probably not due to fiber breakage, but due to partially reversible interface degradation. The interface could be that between fiber and matrix. Damage was observed within the elastic regime, even in the absence of a change in modulus.

Carbon fiber-reinforced concrete can monitor both strain (Chapter 3) and damage simultaneously through electrical resistance measurement. The resistance decreases upon compressive strain and increases upon damage. This means that the strain/stress condition (during dynamic loading) under which damage occurs can be obtained, thus facilitating damage origin identification.

Figure 4.19 [60] shows the fractional change in resistance, strain and stress during repeated compressive loading of carbon fiber (2% by weight of cement)-reinforced concrete at increasing stress amplitudes up to 20% of the compressive strength (within the elastic regime) [60]. The strain returns to zero at the end of each loading cycle. The resistance decreases reversibly upon loading in each cycle. The higher the stress amplitude, the greater is the extent of resistance decrease. As load cycling progresses, the resistance at zero load decreases gradually cycle by cycle. In addition, an extra peak in the resistance curve appears after the first 16 cycles in Fig. 4.19 and becomes larger and larger as cycling progresses. The maximum of the extra peak occurs at the maximum stress of the cycle.

Figure 4.20 [60] shows the fractional change in resistance, strain and stress during repeated compressive loading at increasing and decreasing stress amplitudes. The highest stress amplitude is 40% of the compressive strength. A group of cycles in which the stress amplitude increased cycle by cycle and then decreased cycle by cycle back to the initial low stress amplitude is hereby referred to as a group. Figure 4.20 [60] shows the results for two groups, plus the beginning of the third group. The strain returns to zero at the end of each cycle for any of the stress amplitudes, indicating elastic behavior. Figure 4.21

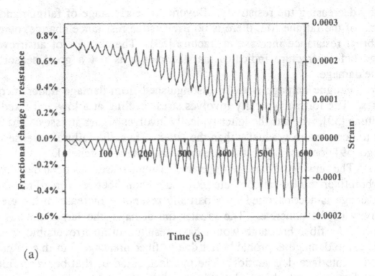

(a)

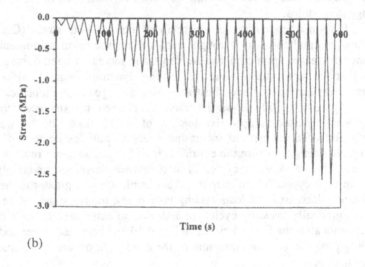

(b)

Fig. 4.19   Fractional change in (a) resistance, (a) strain and (b) stress during repeated compressive loading of carbon fiber-reinforced concrete at increasing stress amplitudes up to 20% of the compressive strength.

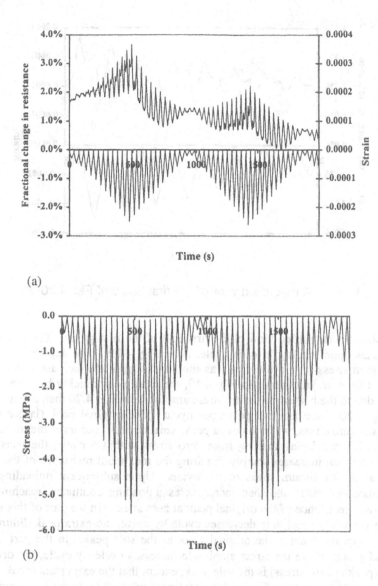

(a)

(b)

Fig. 4.20   Fractional change in (a) resistance, (a) strain and (b) stress during repeated compressive loading of carbon fiber-reinforced concrete at increasing and decreasing stress amplitudes, the highest of which was 40% of the compressive strength.

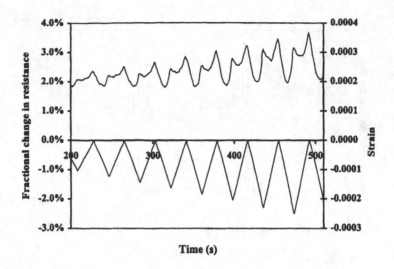

Fig. 4.21   A magnified view of the first 500 s of Fig. 4.20(a).

[60] shows a magnified view of the first half of the first group. The resistance decreases upon loading in each cycle, as in Fig. 4.19. The extra peak at the maximum stress of a cycle grows as the stress amplitude increases, as in Fig. 4.19. However, in contrast to Fig. 4.19, the extra peak quickly becomes quite large, due to the higher maximum stress amplitude in Fig. 4.20 than in Fig. 4.19. In Fig. 4.20, there are two peaks per cycle. The original peak (larger peak) occurs at zero stress, while the extra peak (smaller peak) occurs at the maximum stress. Hence, during loading from zero stress within a cycle, the resistance drops and then increases sharply, reaching the maximum resistance of the extra peak at the maximum stress of the cycle. Upon subsequent unloading, the resistance decreases and then increases as unloading continues, reaching the maximum resistance of the original peak at zero stress. In the part of this group where the stress amplitude decreases cycle by cycle, the extra peak diminishes and disappears, leaving the original peak as the sole peak. In the part of the second group where the stress amplitude increases cycle by cycle, the original peak (peak at zero stress) is the sole peak, except that the extra peak (peak at the maximum stress) returns in a minor way (more minor than in the first group) as the stress amplitude increases. The extra peak grows as the stress amplitude increases, but, in the part of the second group in which the stress amplitude decreases cycle by cycle, it quickly diminishes and vanishes, as in the first group. Within each group, the amplitude of resistance variation increases as the stress amplitude increases and decreases as the stress amplitude subsequently

decreases. The baseline resistance decreases gradually from the first group to the second group.

Fig. 2.37 [60] shows similar results for three successive groups with the highest stress amplitude being 60% of the compressive strength. As the stress amplitude increases, the extra peak at the maximum stress of a cycle grows to the extent that it is comparable to the original peak at zero stress. The decrease of the baseline resistance from group to group is negligible, in contrast to Fig. 4.20. Other features of Figs. 4.20 and 2.37 are similar.

Fig. 4.22 [60] shows four successive groups and the beginning of the fifth group, with the highest stress amplitude being more than 90% of the compressive strength. The highest stress amplitude is the same for each group (Fig. 4.22(b)), but the highest strain amplitude of a group increases from group to group as load cycling progresses (Fig. 4.22(a)). In contrast, the highest strain amplitude of a group does not change from group to group in Fig. 2.37(a). This means that the modulus decreases as cycling occurs in Fig. 4.22, whereas the modulus does not change in Fig. 2.37. In Fig. 4.22, the resistance increases in every cycle. The extra peak at the maximum stress of a cycle is the sole peak in each cycle. The original peak at zero stress does not appear at all. In each group, the amplitude of resistance change in a cycle increases with increasing stress amplitude and subsequently decreases with decreasing stress amplitude. In each group, the resistance increases abruptly as the maximum stress amplitude of the group is about to be reached. The baseline resistance increases gradually from group to group.

Fig. 4.23 [60] shows the results for loading in which the stress amplitude increases cycle by cycle to a maximum (more than 90% of the compressive strength) and is held at the maximum for numerous cycles (Fig. 4.23(b)). The strain amplitude (Fig. 4.23(a)) increases along with the stress amplitude, but continues to increase after the stress amplitude has reached its maximum. This indicates a continuous decrease in modulus after the maximum stress amplitude has been reached. The resistance increases as the stress increases in each cycle, as in Fig. 4.22. The baseline resistance increases significantly cycle by cycle and continues to increase after the stress amplitude has reached its maximum.

Carbon fiber-reinforced concrete is able to sense its own damage, which occurs under increasing stress even within the elastic regime. The damage is partially reversible, as indicated by the partially reversible increase in electrical resistivity observed during cyclic loading at a stress amplitude which increases cycle by cycle. In contrast, compressive strain is indicated by a reversible decrease in resistivity (Chapter 3). Upon increasing the stress, the group in which the stress amplitude increases cycle by cycle. This resistance increase indicates the occurrence of damage. Upon decreasing the stress amplitude, the extra peak does not occur, except for the first two cycles of stress amplitude decrease. The greater the stress amplitude, the larger and the less

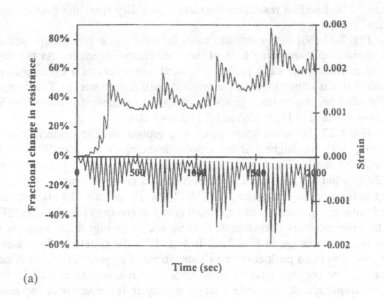

(a)

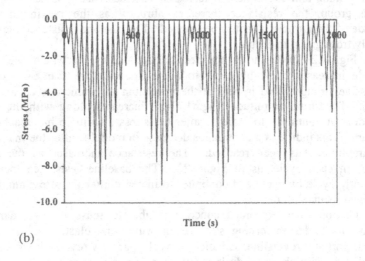

(b)

Fig. 4.22   Fractional change in (a) resistance, (a) strain and (b) stress during repeated compressive loading of carbon fiber-reinforced concrete at increasing and decreasing stress amplitudes, the highest of which was >90% of the compressive strength.

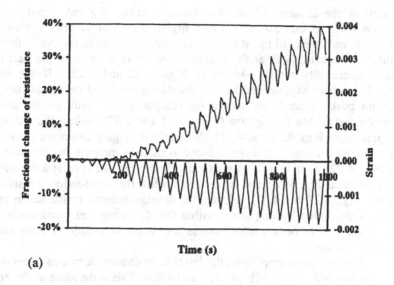

(a)

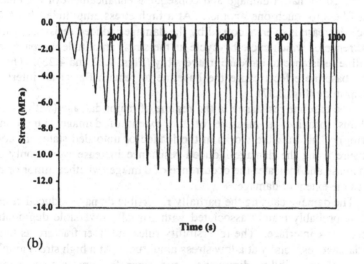

(b)

Fig. 4.23 Fractional change in (a) resistance, (a) strain and (b) stress during repeated compressive loading of carbon fiber-reinforced concrete at increasing stress amplitudes up to >90% of the compressive strength and then with the stress amplitude fixed at the maximum.

reversible is the damage-induced resistance increase (the extra peak). The resistance starts to increase at a stress higher than that in prior cycles and continues to increase until the stress reaches the maximum in the cycle, thereby resulting in the extra peak at the maximum stress of a cycle in the part of a partial irreversibility clearly shown in Figs. 4.22 and 4.23. If the stress amplitude has been experienced before, the damage-induced resistance increase (the extra peak) is small, as shown by comparing the result of the second groupwith that of the first group (Figs. 4.20 and 2.37), unless the extent of damage is large (Figs. 4.22 and 4.23). When the damage is extensive (as shown by a modulus decrease), damage-induced resistance increase occurs in every cycle (Fig. 4.22), even at a fixed stress amplitude (Fig. 4.23) or at a decreasing stress amplitude (Fig. 4.22), and it can overshadow the strain-induced resistance decrease (Figs. 4.22 and 4.23). Hence, the damage-induced resistance increase occurs mainly during loading (even within the elastic regime), particularly at a stress above that in prior cycles, unless the stress amplitude is high and/or damage is extensive.

At a low stress amplitude, the baseline resistance decreases irreversibly and gradually cycle by cycle (Figs. 4.19 and 4.20). This is the same as the effect [58] attributed to matrix damage and consequent enhancement of the chance of adjacent fibers to touch one another. At a high stress amplitude, this baseline resistance decrease is overshadowed by the damage-induced resistance increase, the occurrence of which cycle-by-cycle as the stress amplitude increases causes the baseline resistance to increase irreversibly (Figs. 4.22 and 4.23). These two opposing baseline effects cause the baseline to remain flat at an intermediate stress amplitude (Fig. 2.37).

The baseline resistance in the regime of major damage (with a decrease in modulus) provides a measure of the extent of damage (i.e., condition monitoring). This measure works in the loaded or unloaded state. In contrast, the measure using the damage-induced resistance increase works only during stress increase and indicates the occurrence of damage (whether minor or major) as well as the extent of damage.

The damage causing the partially reversible damage-induced resistance increase is probably mainly associated with partially reversible degradation of the fiber-matrix interface. The reversibility rules out fiber fracture as the main type of damage, especially at a low stress amplitude. At a high stress amplitude, the extent of reversibility diminishes, and fiber fracture may contribute to causing the damage. Fiber fracture can occur during the opening of a crack that is bridged by a fiber. The fiber-matrix interface degradation may be associated with slight fiber pull-out upon slight crack opening for cracks that are bridged by fibers. The severity of the damage-induced resistance increase supports the involvement of the fibers in the damage mechanism, as the fibers are much more conducting than the matrix.

In the regime of elastic deformation, the damage does not affect the strain permanently, as shown by the total reversibility of the strain during cyclic loading (Figs. 4.19-4.23 and 2.37). Nevertheless, damage occurs during stress increase, as shown by the damage-induced resistance increase. Damage occurs even in the absence of a change in modulus. Hence, the damage-induced resistance increase is a sensitive indicator of minor damage (without a change in modulus), in addition to being a sensitive indicator of major damage (with a decrease in modulus). In contrast, the baseline resistance increase is an indicator of major damage only.

4.3.3   Sensing damage at the interface between concrete and steel rebar

Steel-reinforced concrete is a widely used structural material. The effectiveness of the steel reinforcement depends on the bond between the steel reinforcing bar (rebar) and the concrete. Destructive measurement of the shear bond strength by pull-out, push-in and related testing methods is commonly used to assess the quality of the bond [62-76]. Nondestructive methods of bond assessment are attractive for condition evaluation in the field. They include acoustic [77-79] and electrical [80] methods. In particular, measurement of the contact electrical resistivity of the bond interface has been used to investigate

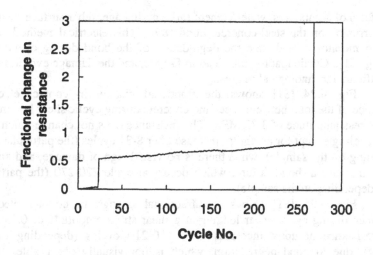

Fig. 4.24   Variation of the fractional contact resistance change ($\Delta R/R_o$) with cycle no. during cyclic shear loading at a shear stress amplitude of 3.73 MPa up to bond failure. The contact resistance is that of the interface between concrete and steel rebar.

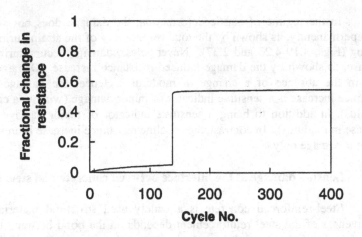

Fig. 4.25   Variation of the fractional contact resistance change
            ($\Delta R/R_o$) with cycle no. during cyclic shear loading at a
            shear stress amplitude of 0.75 MPa.   The test was
            stopped prior to bond failure.   The contact resistance is
            that of the interface between concrete and steel rebar.

the effects of admixtures, water/cement ratio, curing age, rebar surface treatment
and corrosion on the steel-concrete bond [80].   This electrical method can be
used to monitor in real time the degradation of the bond during cyclic shear
loading [81]. Cyclic loading may lead to fatigue, and the damage evolution is of
scientific and technological interest.

Fig. 4.24 [81] shows the fractional change in contact electrical
resistance of the joint between steel and concrete during cyclic shear loading at a
shear stress amplitude of 3.73 MPa.   The resistance does not change much upon
stress cycling except for an abrupt increase after 8-31 cycles (the particular cycle
depending on the sample), when there is no visual sign of damage, and another
abrupt increase at bond failure, which occurs at cycle 220-270 (the particular
cycle depending on the sample).

Fig. 4.25 [81] shows the fractional change in contact electrical
resistance during cyclic shear loading at a shear stress amplitude of 0.75 MPa.
The resistance abruptly increases after 150-210 cycles (depending on the
sample), due to bond degradation, which is not visually observable.   Bond
failure does not occur up to 400 cycles, at which testing is stopped. The bond
strength before any cyclic shear is 6.68 ± 0.24 MPa, and after the abrupt increase
(at the end of 400 cycles in Fig. 4.25) is 5.54 ± 0.43 MPa. Thus, even though
the abrupt increase does not cause visually observable damage, bond
degradation occurs.

Comparison of Figs. 4.24 and 4.25 shows that a higher stress amplitude causes bond degradation and bond failure to occur at lower numbers of cycles, as expected.

The abrupt increase in resistance due to bond degradation (not bond failure) (Figs. 4.24 and 4.25) provides a method of monitoring bond quality nondestructively in real time during dynamic loading. In contrast, bond strength measurement by mechanical testing is destructive. The bond degradation is attributed to fatigue.

4.3.4 Sensing damage at the interface between new concrete and old concrete

The repair of a concrete structure commonly involves the bonding of new concrete to the old concrete [77-88]. Partly due to the drying shrinkage of the new concrete, the quality of the bond is limited. Destructive measurement of the shear bond strength has been previously used to assess the quality of the bond [89]. However, the bond may degrade at stresses below the shear bond strength, even though the degradation may not be visible. This degradation may occur during static or cyclic loading. In particular, cyclic loading may lead to fatigue. Such degradation is revealed by measurement of the contact electrical

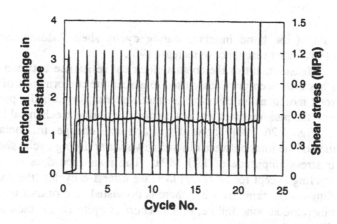

Fig. 4.26 Variation of the fractional contact resistance change with cycle no. during cyclic shear loading of a joint between old and new mortar at a shear stress amplitude of 1.21 MPa up to bond failure. Thick curve: fractional change in contact resistance. Thin curve: shear stress. The contact resistance is that of the interface between new mortar and old mortar.

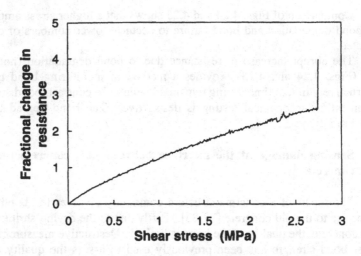

Fig. 4.27   Variation of the fractional contact resistance change with
                      shear stress during static shear loading of a joint
                      between old and new mortar up to failure. The contact
                      resistance is that of the interface between new mortar
                      and old mortar.

resistance of the bond interface during cyclic shear loading, as degradation causes the contact resistance to increase [90].

Measurement of the contact electrical resistance between old and new mortar has also been previously used to assess the performance of carbon fiber reinforced mortar as an electrical contact material for cathodic protection [91]. However, the measurement was not carried out during mechanical loading.

Fig. 4.26 [90] shows the fractional change in contact electrical resistance of the joint between old and new mortar during cyclic shear loading at a shear stress amplitude of 1.21 MPa. The resistance does not change upon stress cycling except for an abrupt increase after 1-6 cycles (the particular cycle depending on the sample), when there is no visual sign of damage, and another abrupt increase at bond failure, which occurs at cycle 18-27 (the particular cycle depending on the sample).

The bond strength before the first abrupt increase is 2.87 ± 0.18 MPa; that after the first abrupt increase is 2.38 ± 0.22 MPa. Thus, even though the first abrupt increase does not cause visually observable damage, bond degradation occurs.

During static loading, the contact resistance increases monotonically with increasing shear stress and abruptly increases at bond failure, as shown in Fig. 4.27 [90] for the case of a specimen which has not been loaded prior to the measurement. No abrupt increase in resistance occurs during static loading prior to failure, in contrast to the observation of an abrupt increase prior to fatigue failure (Fig. 4.26).

Fig. 4.28 [90] shows the fractional change in contact electrical resistance during cyclic shear loading at a shear stress amplitude of 0.97 MPa (lower than that of Fig. 4.26). The resistance showes the first abrupt increase after 22-48 cycles (the particular cycle depending on the sample), and another abrupt increase at bond failure, which occurs after 69-92 cycles (the particular cycle depending on the sample).

Fig. 4.29 [90] shows the fractional change in contact electrical resistance during cyclic shear loading at a shear stress amplitude of 0.81 MPa (lower than that of Fig. 4.27). The resistance abruptly increases after 557-690 cycles (depending on the sample), due to bond degradation, which is not visually observable. Bond failure did not occur up to 1300 cycles, at which testing was stopped.

Comparison of Figs. 4.26, 4.28 and 4.29 shows that a higher stress amplitude causes bond degradation and bond failure to occur at lower numbers of cycles, as expected.

The abrupt increase in resistance due to bond degradation (not bond failure) (Figs. 4.26, 4.28 and 4.29) provides a method of monitoring bond quality nondestructively in real time during dynamic loading. In contrast, bond

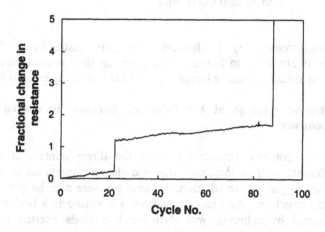

Fig. 4.28   Variation of the fractional contact resistance change with cycle no. during cyclic shear loading of a joint between old and new mortar at a shear stress amplitude of 0.97 MPa up to bond failure. The contact resistance is that of the interface between new mortar and old mortar.

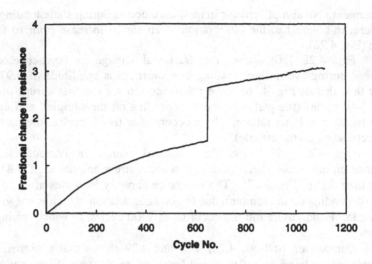

Fig. 4.29   Variation of the fractional contact resistance change with
cycle no. during cyclic shear loading of a joint between
old and new mortar at a shear stress amplitude of 0.81
MPa. The test was stopped prior to bond failure. The
contact resistance is that of the interface between new
mortar and old mortar.

strength measurement by mechanical testing is destructive.   The bond
degradation is attributed to fatigue. This interpretation is consistent with the
absence of an abrupt resistance increase during static loading prior to failure.

4.3.5   Sensing damage at the interface between unbonded concrete elements

Many concrete structures involve the direct contact of one cured
concrete element with another, such that one element exerts static pressure on
the other due to gravity. In addition, dynamic pressure may be exerted by live
loads on the structure. An example of such a structure is a bridge involving
slabs supported by columns, with dynamic live loads exerted by vehicles
traveling on the bridge. Another example is a concrete floor in the form of slabs
supported by columns, with live loads exerted by people walking on the floor.
The interface between concrete elements that are in pressure contact is of
interest, as it affects the integrity and reliability of the assembly. For example,
deformation at the interface affects the interfacial structure, which can affect the
effectiveness of load transfer between the contacting elements and can affect the

durability of the interface to the environment. Moreover, deformation at the interface can affect the dimensional stability of the assembly. Of particular concern is how the interface is affected by dynamic loads.

Effective study of the interface between concrete elements that are in pressure contact and under dynamic loading requires the monitoring of the interface during dynamic loading. Hence, a nondestructive monitoring technique that provides information in real time during dynamic loading is desirable. Microscopic examination of the interface viewed at the edge cannot effectively provide interfacial information, though it can be nondestructive and be in real time. Microscopic examination of the interface surfaces after separation of the contacting elements can provide microstructural information, but it cannot be performed in real time. Mechanical testing of the interface, say under shear, can provide interfacial information, but it is destructive (unless the shear strain amplitude is within the elastic regime) and it cannot be conveniently performed in real time (due to the difficulty of having simultaneous dynamic compression and dynamic shear). The difficulties and ineffectiveness associated with these conventional techniques contribute to the scarcity of work on concrete-concrete pressure contacts.

In this section, contact electrical resistance measurement is used to monitor concrete-concrete pressure contacts in real time during dynamic pressure application. As the surface of concrete is never perfectly smooth, asperities occur on the surface, thus causing the true contact area at the interface to be much smaller than the geometric junction area. As a consequence, the local stress at the asperities is much higher than the overall stress applied to the junction. The greater the true contact area, the lower the contact resistance. Deformation (flattening) of the asperities, as caused by the high local stress at the asperities, increases the true contact area. Therefore, the interfacial structure is changed. The contact resistance provides information on the interfacial structure, particularly in relation to the deformation at the interface. By monitoring the contact resistance in real time during loading and unloading, the extent, reversibility and loading history dependence of the deformation at various points of loading and unloading can be investigated, thus providing information on the structure and dynamic behavior of the interface.

Since concrete is somewhat conductive electrically, the contact resistance of the interface between contacting concrete elements can be conveniently measured by using the concrete elements as electrical leads – two for passing current and two for voltage measurement (i.e., the four-probe method), as provided by two concrete beams that overlap at 90° (Fig. 4.30). The volume resistance of each lead is negligible compared to the contact resistance of the junction, so the measured resistance (i.e., voltage divided by current) is the contact resistance. The contact resistance multiplied by the junction area gives the contact resistivity, which is independent of the junction area and describes the structure of the interface.

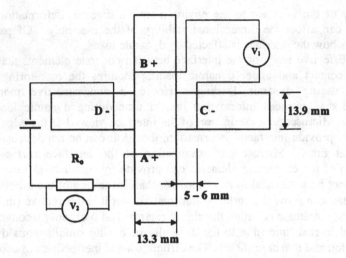

Fig. 4.30   Sample configuration for measurement of the contact electrical resistance of the interface between unbonded mortar elements.

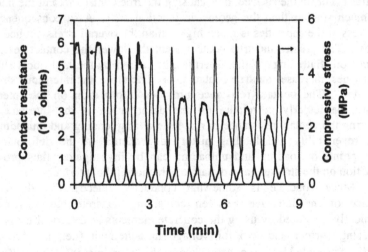

Fig. 4.31   Variation of contact resistance with time and of compressive stress with time during cyclic compression at a stress amplitude of 5 MPa. The contact resistance is that of the interface between unbonded mortar elements.

The data below [91] involves the use of mortar (with fine aggregate but not coarse aggregate) instead of concrete (with both fine and coarse aggregates). However, the interfacial effects should be quite similar for mortar and concrete.

Fig. 4.31 shows the variation in resistance and stress during cyclic compressive loading at a stress amplitude of 5.0 MPa. The compressive strength of the mortar used is 64 ± 2 MPa, as determined by compressive testing of 51 x 51 x 51 mm (2 x 2 x 2 in) cubes. The stress-strain curve is a straight line up to failure. In every cycle, the resistance decreases as the compressive stress increases, such that the maximum stress corresponds to the minimum resistance and the minimum stress (zero stress) corresponds to the maximum resistance. The minimum resistance (at the maximum stress) increases slightly as cycling progresses, but the maximum resistance (at the minimum or zero stress) decreases with cycling. Due to the asperities at the interface, the local compressive stress on the asperities is much higher than the overall compressive stress. As a result, plastic deformation occurs at the asperities, which means that more contact area is created during cycling. The occurrence of deformation is supported by the crosshead displacement observed within each cycle. The displacement is greatest (i.e., most deformation) at the maximum stress within each cycle and is not totally reversible. The plastic deformation is why the observed electrical resistance at the minimum stress (i.e., upon unloading)

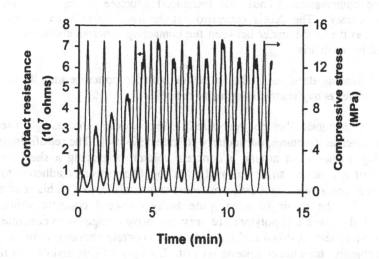

Fig. 4.32   Variation of contact resistance with time and of compressive stress with time during cyclic compression at a stress amplitude of 15 MPa. The contact resistance is that of the interface between unbonded mortar elements.

decreases as cycling progresses. On the other hand, due to the brittleness of the mortar, the compressive loading probably causes fracture at some of the asperities, thereby generating debris, which increases the contact resistance. Debris generation is probably the reason for the slight increase in the contact resistance at the maximum stress as cycling progresses. After about seven loading cycles, the maximum resistance (at the minimum stress) levels off, due to the limit of the extent of flattening of the asperites. However, the slight increase of the minimum resistance (at the maximum stress) persists beyond the first seven cycles, probably due to the continued generation of debris as cycling progresses.

The stress amplitude in Fig. 4.32 is 15.0 MPa, which is higher than that in Fig. 4.31. The minimum resistance (at the maximum stress) increases with cycling more significantly than in Fig. 4.31. This is probably due to the more significant debris generation at the higher stress amplitude. The maximum resistance (at the minimum stress) increases in the first four cycles. This is probably due to the effect of debris generation overshadowing the effect of the flattening of the asperities. After four cycles, the maximum resistance essentially levels off, probably due to the limit of the extent of debris generation for this stress amplitude.

The results above mean that, even at a low compressive stress amplitude of 5 MPa, the structure of a concrete-concrete contact changes during dynamic compression. Thus, the interfacial structure is dependent on the loading history. The debris generation at the interface may be of practical concern, as the load transfer between the contacting concrete elements may be affected by the debris.

4.3.6 Sensing damage at the interface between concrete and its carbon fiber epoxy matrix composite retrofit

Continuous fiber polymer-matrix composites are increasingly used to retrofit concrete structures, particularly columns [92-104]. The retrofit involves wrapping a fiber sheet around a concrete column or placing a sheet on the surface of a concrete structure, such that the fiber sheet is adhered to the underlying concrete using a polymer, most commonly epoxy. This method is effective for the repair of even quite badly damaged concrete structures. Although the fibers and polymer are very expensive compared to concrete, the alternative of tearing down and rebuilding the concrete structure is often even more expensive than the composite retrofit. Both glass fibers and carbon fibers are used for the composite retrofit. Glass fibers are advantageous for their relatively low cost, but carbon fibers are advantageous for their high tensile modulus.

The effectiveness of a composite retrofit depends on the quality of the bond between the composite and the underlying concrete, as good bonding is necessary for load transfer. Peel testing for bond quality evaluation is destructive [105]. Nondestructive methods to evaluate the bond quality are valuable. They include acoustic methods, which are not sensitive to small amounts of debonding or bond degradation [106], and dynamic mechanical testing [107]. This section uses electrical resistance measurement for nondestructive evaluation of the interface between concrete and its carbon fiber composite retrofit [108]. The method is effective for studying the effect of debonding stress on the interface. The concept behind the method is that bond degradation causes the electrical contact between the carbon fiber composite retrofit and the underlying concrete to degrade. Since concrete is electrically more conductive than air, the presence of an air pocket at the interface causes the measured apparent volume resistance of the composite retrofit in a direction in the plane of the interface to increase. Hence, bond degradation is accompanied by an increase in the apparent resistance of the composite retrofit. Although the polymer matrix (epoxy) is electrically insulating, the presence of a thin layer of epoxy at the interface was found to be unable to electrically isolate the composite retrofit from the underlying concrete.

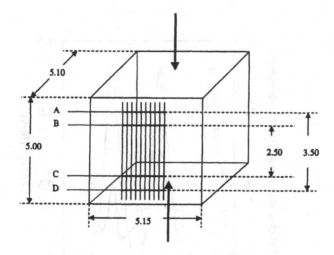

Fig. 4.33   Sample configuration. The vertical arrow indicates the direction of compressive loading. All dimensions are in cm. A, B, C and D are the four electrical leads emanating from the four electrical contacts (thick horizontal lines), which are attached to the fiber retrofit indicated by vertical parallel lines on the front face of the concrete block. The fiber direction is in the stress direction (vertical).

A 40 x 15 mm sample of carbon fiber sheet (the composite retrofit), with the fibers along the 40 mm length of the sample, is pressed against a surface of the polished concrete block while the epoxy resin is at the interface for the purpose of bonding the fiber sheet to the concrete, as illustrated in Fig. 4.33 [108]. The curing of the epoxy resin is carried out at room temperature.

Four electrical contacts (A, B, C and D) are applied at four points along the 40 mm length of the fiber sheet sample, such that each contact is a strip stretching across the 15 mm width of the sample (Fig. 4.33). Each electrical contact is in the form of silver paint in conjunction with copper wire. In the four-probe method used for DC electrical resistance measurement, two of the electrical contacts (A and D, Fig. 4.33) are for passing current; the remaining two contacts (B and C) are for measuring voltage. The voltage divided by the current gives the measured resistance, which is the apparent volume resistance of the fiber sheet between B and C when the sheet is in contact with the concrete substrate.

Uniaxial compression is applied on a concrete block with the fiber sheet on one surface, such that the stress is in the fiber direction (Fig. 4.33), while the electrical resistance is continuously measured.

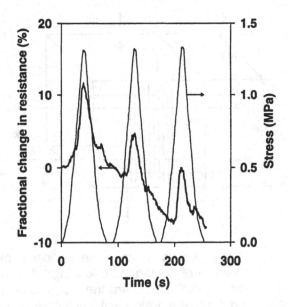

Fig. 4.34   The fractional change in resistance for the fiber retrofit on a concrete substrate during cyclic compressive loading.

Fig. 4.34 shows the fractional change in resistance during cyclic compressive loading at a stress amplitude of 1.3 MPa. The stress is along the fiber direction. Stress returns to zero at the end of each cycle. In each cycle, the electrical resistance increases reversibly during compressive loading. This is attributed to the reversible degradation of the bond between carbon fiber sheet and concrete substrate during compressive loading. This bond degradation decreases the chance for fibers to touch the concrete substrate, thereby leading to a resistance increase.

As cycling progresses, both the maximum and minimum values of the fractional change in resistance in a cycle decrease. This is attributed to the irreversible disturbance in the fiber arrangement during repeated loading and unloading. This disturbance increases the chance for fibers to touch the concrete substrate, thereby causing the resistance to decrease irreversibly as cycling progresses.

As shown in Fig. 4.34, the first cycle exhibits the highest value of the fractional change in resistance. This is due to the greatest extent of bond degradation taking place during the first cycle.

4.4 Sensing damage due to freeze-thaw cycling in a cement-based material

Freeze-thaw cycling is one of the main causes of degradation of concrete in cold regions. The degradation stems from the freezing of the water in the concrete upon cooling, and the thawing upon subsequent heating. The phase transition is accompanied by dimensional change and internal stress change. Freeze-thaw cycling can result in failure.

Research on the freeze-thaw durability of cement-based materials has been focused on the mechanical property degradation (e.g., modulus and strength) [109-111], weight change [109,111-113], length change [113,114], microstructural change [115] and ultrasonic signature change [113,116] after different amounts of freeze-thaw cycling. Relatively little attention has been previously given to monitoring during freeze-thaw cycling. Techniques previously used for real-time monitoring include strain measurement [114] and electrical resistivity measurement [117]. Without real-time monitoring, the degradation could not be monitored during freeze-thaw cycling. Therefore, study of the damage evolution required testing numerous specimens at different numbers of freeze-thaw cycles. As different specimens are bound to be a little different in the degree of perfection, the testing of different specimens gives data scatter which makes it difficult to study the damage evolution. In order to study the damage evolution on a single specimen during freeze-thaw cycling, a nondestructive and sensitive real-time testing method is necessary.

Electrical resistivity measurement is a nondestructive method. The electrical resistivity of cement paste decreases reversibly upon heating at temperature above 0°C (without freezing or thawing), due to the existence of an activation energy for electrical conduction [118]. This phenomenon allows cement paste to function as a thermistor for sensing temperature. Thus, electrical resistivity measurement allows simultaneous monitoring of both temperature and damage. A temperature increase causes the resistivity to decrease reversibly, whereas damage causes the resistivity to increase irreversibly.

Fig. 4.35 [119] shows the fractional change in resistivity and the temperature during fast thermal cycling (40 min per cycle) of mortar (without fiber) between –20 and 52°C. The resistivity decreases upon heating and increases upon cooling in every cycle, due to the existence of an activation energy for electrical conduction. The resistivity changes smoothly and similarly above and below 0°C, indicating that the phase transition does not affect the resistivity. Even when the heating and cooling rates are very low, i.e., 6 h per cycle, the phase transition at 0°C has only slight effect on the resistivity (Fig. 4.36). Although the resistivity changes abruptly at 0°C, the effects of freezing and thawing on the resistivity are small compared to the effect of temperature on the resistivity. It is reasonable that this small effect is only observed when the heating and cooling rates are low.

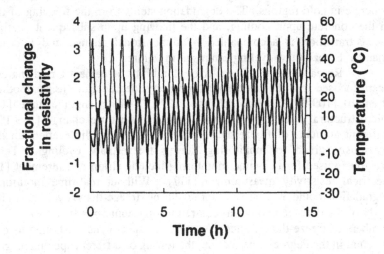

Fig. 4.35   The fractional change in resistivity vs. time (thick curve) and the temperature vs. time (thin curve) during fast freeze-thaw cycling (40 min per cycle) of mortar (without fiber).

The resistivity at the end of a heating-cooling cycle is higher than that at the beginning of the cycle (Fig. 4.35). In other words, the upper envelope of the resistivity variation (corresponding to the resistivity at -20°C) increases cycle by cycle. The lower envelope (corresponding to the resistivity at 52°C) also increases cycle by cycle, but the increase is less significant than that of the upper envelope. As a consequence, the amplitude of resistivity variation increases with cycling. This behavior is attributed to damage, which causes the resistivity to increase irreversibly. That the upper envelope upshifts more than the lower envelope means that the damage occurs more significantly upon cooling than upon heating. This is expected since (i) thermal contraction occurs upon cooling and the surface of the specimen cools faster than the center of the specimen and (ii) water expands upon freezing.

Upon freeze-thaw failure, the resistivity rises abruptly essentially to infinity, as observed before the completion of 15 hours of cycling (Fig. 4.35). This rise occurs at -20°C (the coldest point of a cycle), again indicating that damage during cooling is more significant than that during heating. Prior to failure, no abrupt resistivity increase was observed. This means that the damage evolution involves damage accumulating gradually cycle by cycle, until failure occurs.

At a given temperature, the resistivity during heating is slightly lower than that during subsequent cooling, as shown in Fig. 4.36. The hysteresis

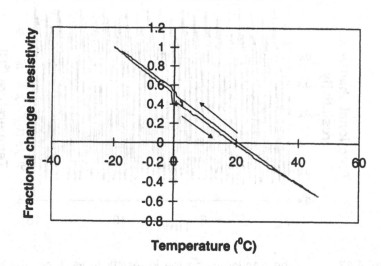

Fig. 4.36   The fractional change in resistivity vs. temperature in a cycle of slow freeze-thaw cycling (6 h per cycle) of mortar (without fiber).

becomes more severe as cycling progresses. The hysteresis is attributed to the damage inflicted during cooling and the association of damage with a higher resistivity. That damage infliction occurs smoothly throughout cooling from 52 to -20°C means that the damage is not due to freezing itself, but is due to thermal contraction and the fact that the surface cools faster than the center of the specimen.

Fig. 4.37 shows the fractional change in resistivity and the temperature during fast thermal cycling (40 min per cycle) between 0 and 52°C (i.e., without freezing). The resistivity decreases reversibly upon heating, due to the existence of an activation energy for electrical conduction. In contrast to the case of freeze-thaw cycling at a similar cycling rate (Fig. 4.35), the lower envelope of resistivity variation does not shift upon cycling and the upper envelope upshifts only slightly. As the irreversible increase in resistivity is associated with damage, this means that the damage during thermal cycling without freezing is negligible compared to that during freeze-thaw cycling. As a result, failure does not occur after 15 h of thermal cycling without freezing, but was visually observed before the end of 15 h of freeze-thaw cycling (Fig. 4.35).

As mentioned above, the damage in Fig. 4.35 was not due to freezing itself, but was due to thermal contraction and the fact that the surface cooled

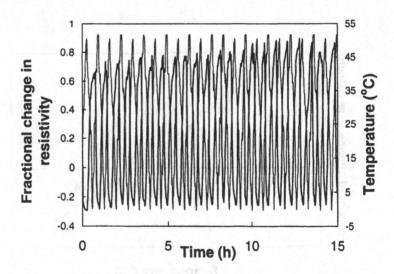

Fig. 4.37   The fractional change in resistivity vs. time (thick curve) and the temperature vs. time (thin curve) during temperature cycling of mortar (without fiber) without freezing.

faster than the center of the specimen. Comparison between Fig. 4.35 and Fig. 4.37 shows that the damage caused by thermal contraction is significant in the presence of freezing, but negligible in the absence of freezing. In other words, freezing aggravates the damage that is due to thermal contraction.

The thermal damage observed in Fig. 4.35 is not related to damage that occurs at elevated temperatures (up to 52°C). This is shown by a separate experiment in which the resistivity was monitored over time up to 4000 s at a constant temperature of 50°C. The resistivity was observed to increase by less than 2%, in contrast to the much larger fractional increase in resistivity (whether the upper envelope or the lower envelope) in Fig. 4.35.

4.5 Sensing damage due to drying shrinkage in a cement-based material

The hydration reaction that occurs during the curing of cement causes shrinkage, called autogenous shrinkage, which is accompanied by a decrease in the relative humidity within the pores. When the curing is conducted in an open atmosphere, as is usually the case, additional shrinkage occurs due to the movement of water through the pores to the surface and the loss of water on the surface by evaporation. The overall shrinkage that occurs in this case is known as the drying shrinkage, which is the shrinkage that is practically important.

The shrinkage of cement-based materials during curing is a cause of defects (such as cracks) in cement-based materials. It can also cause prestressing loss [120]. The tendency for defect formation during shrinkage increases with increasing size of the cement-based material. Thus, the problem is particularly serious for large concrete structures such as floors and dams.

The effects of shrinkage have been studied by numerous workers by measurement of the shrinkage strain and observation of the cracks. However, the microstructural change, which necessarily proceeds the cracking, has not received much attention. The extent of microstructural change and the evolution of the microstructure as shrinkage occurs are important for the understanding of the shrinkage process. This understanding is valuable for the alleviation of the problem associated with shrinkage induced cracking.

Silica fume [121-124] is very fine noncrystalline silica produced by electric arc furnaces as a by-product of the production of metallic silicon or ferrosilicon alloys. It is a powder with particles having diameters 100 times smaller than those of anhydrous portland cement particles, i.e., mean particle size between 0.1 and 0.2 μm. The SiO_2 content ranges from 85 to 98%. Silica fume is pozzolanic.

Silica fume used as an admixture in a concrete mix has significant effects on the properties of the resulting material [125]. These effects pertain to the strength, modulus, ductility, vibration damping capacity, sound absorption, abrasion resistance, air void content, shrinkage, bonding strength with reinforcing steel, permeability, chemical attack resistance, alkali-silica reactivity reduction, corrosion resistance of embedded steel reinforcement, freeze-thaw durability, creep rate, coefficient of thermal expansion, specific heat, thermal conductivity, defect dynamics, dielectric constant, and degree of fiber dispersion in mixes containing short microfibers. In addition, silica fume addition degrades the workability of the mix.

The addition of untreated silica fume to cement paste decreases the drying shrinkage [120,126-131]. This desirable effect is partly due to the reduction of the pore size and connectivity of the voids and partly due to the prestressing effect of silica fume, which restrains the shrinkage. The use of silane-treated silica fume in place of untreated silica fume further decreases the drying shrinkage, due to the hydrophylic character of the silane-treated silica fume and the formation of chemical bonds between silica fume particles and cement [120]. The use of silane and untreated silica fume as two admixtures also decreases the drying shrinkage, but not as significantly as the use of silane-treated silica fume [120]. However, silica fume has also been reported to increase the drying shrinkage [121,132,133], and the restrained shrinkage crack width is increased by silica fume addition [134].

Due to the pozzolanic nature of silica fume, silica fume addition increases the autogenous shrinkage, as well as the autogenous relative humidity change [135,136]. These effects are undesirable, as they may cause cracking if the deformation is restrained. Aggregates are known to decrease the drying shrinkage [137].

In order to study the effects of silica fume and fine aggregate on the shrinkage-induced microstructural change, this section addresses the drying shrinkage of cement pastes with and without silica fume (untreated) and of mortar without silica fume. Both the shrinkage strain and the electrical resistivity (related to the microstructure and obtained from the electrical resistance and the strain) were measured continually from 1 day to 28 days of curing [138].

Table 4.2   Volume electrical resistivity and resistance at 1 day of curing

Material	Resistivity (Ω.cm)	Resistance ($M\Omega$)	Water/cement ratio
*(a) Plain	1.01×10^6	0.233	0.30
*(b) Plain	1.06×10^6	0.244	0.35
*(c) Plain	1.11×10^6	0.257	0.40
*(d) With silica fume	5.46×10^5	0.126	0.35
†(e) With sand	1.56×10^7	3.59	0.35

* Cement paste
† Mortar

Table 4.2 shows the initial (1 day of curing) values of the volume electrical resistivity and resistance for each of the five compositions investigated [138]. The resistivity increases with increasing water/cement ratio. It is decreased by the addition of silica fume and is increased by the addition of sand.

Figs. 4.38 and 4.39 show the shrinkage strain and fractional change in resistivity respectively vs. curing age. The presence of silica fume decreases both shrinkage strain and fractional change in resistivity at the same curing age for all curing ages from 1 to 28 days. This means that the silica fume restrains the drying shrinkage as well as the shrinkage-induced microstructural change. Both shrinkage strain and fractional change in resistivity increase smoothly with increasing curing age, such that the increase becomes more gradual as curing progresses.

Sand decreases the shrinkage strain even more than silica fume (Fig. 4.38), but the fractional change in resistivity is increased by sand. This means that the shrinkage-induced microstructural change is larger when sand is present, presumably due to the effect of shrinkage on the microstructure of the interface between sand and cement. Sand does not shrink while cement shrinks, thereby resulting in microstructural changes at the sand-cement interface as drying shrinkage proceeds. The interface is associated with a contact electrical resistance, which increases as the interfacial voids or void precursors become

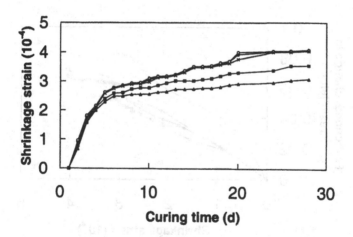

Fig. 4.38   Shrinkage strain vs. curing time for plain cement paste with water/cement ratio = 0.30 (x), plain cement paste with water/cement ratio = 0.35 (•), plain cement paste with water/cement ratio = 0.40 (o), silica fume cement paste (■), and plain mortar (▲).

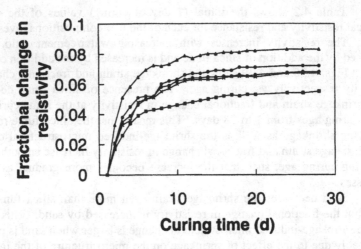

Fig. 4.39   Fractional change in resistivity vs. curing time for plain
cement paste with water/cement ratio = 0.30 (×), plain
cement paste with water/cement ratio = 0.35 (•), plain
cement paste with water/cement ratio = 0.40 (o), silica
fume cement paste (■), and plain mortar (▲).

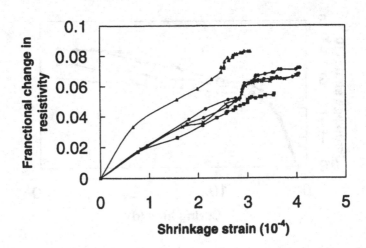

Fig. 4.40   Fractional change in resistivity vs. shrinkage strain for
plain cement paste with water/cement ratio = 0.30 (×),
plain cement paste with water/cement ratio = 0.35 (•),
plain cement paste with water/cement ratio = 0.40 (o),
silica fume cement paste (■), and plain mortar (▲).

more numerous. The increase of the contact resistivity between steel rebar and concrete as drying shrinkage proceeds has been reported [80].

An increase in the water/cement ratio causes a negligible increase in the shrinkage strain (Fig. 4.38), but a slight increase in the fractional change in resistivity (Fig. 4.39), as shown by comparing the three plain cement pastes with water/cement ratios of 0.30, 0.35 and 0.40. This means that the shrinkage-induced microstructural change increases slightly with increasing water/cement ratio. The effect of the water/cement ratio is much smaller than that of silica fume or sand.

Fig. 4.40 shows that the fractional change in resistivity is less in the presence of silica fume for the same strain. This implies that the extent of microstructural change at the same strain is less in the presence of silica fume. Fig. 4.40 also shows that the fractional change in resistivity is higher in the presence of sand for the same strain. Thus, the extent of microstructural change at the same strain is more in the presence of sand.

Fig. 4.40 shows that the fractional change in resistivity abruptly increases at a strain of 3.0×10^{-4} for all cement pastes without silica fume and at a strain of 2.5×10^{-4} for mortar. This abrupt resistivity increase is probably associated with an abrupt and irreversible microstructural change. The addition of silica fume essentially eliminates this effect, whereas the addition of sand causes the microstructural change to occur at a lower shrinkage strain. This is consistent with the notion that silica fume addition diminishes the shrinkage-induced microstructural change, whereas sand addition increases this quantity.

The fractional change in resistivity per unit strain is in the range from 150 to 500. This quantity describes the severity of shrinkage-induced microstructural change. The severity is slightly lower in the presence of silica fume, and is significantly higher in the presence of sand. The severity tends to decrease as shrinkage proceeds, as expected from the decreasing rate of shrinkage as shrinkage proceeds (Fig. 4.38).

The fractional change in resistivity per unit compressive strain in the cured state, as determined during compressive loading, is 10 [42]. Thus, the microstructural change induced by shrinkage strain is much larger than that induced by compressive strain for the same amount of strain. Nevertheless, both shrinkage strain and compressive strain cause the resistivity in the strain direction to increase. The large microstructural change during drying shrinkage is expected from the hydration reaction which takes place during curing.

4.6 Sensing damage due to creep in a cement-based material

Creep [139,140] is a form of time-dependent plastic deformation that occurs under load, which is typically fixed during creep testing. Creep affects the dimensions of a component, and dimensional stability is important for many structural components, including concrete slabs and columns. Although creep

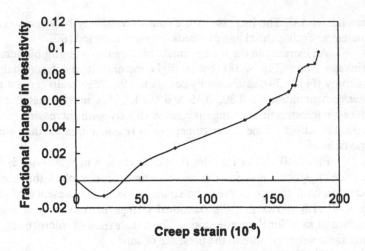

Fig. 4.41   Fractional change in resistivity vs. compressive strain during a month of creep testing of plain cement mortar at a constant compressive stress.

during concrete curing [141-146] is more significant than that after curing [147-151], creep after curing is relevant to the durability and stability of structures during use.   Thus, this section addresses creep after curing.   Creep is more severe at elevated temperatures [152,153], but this section is limited to creep at room temperature.

Research on creep has been focused on the strain during creep [154-158], rather than the material property variation during creep.   The microstructure affects numerous properties, including mechanical and electrical properties.   For the purpose of understanding the microstructural effect of creep, it is desirable to investigate the property variation during creep.   It is further preferred that the property be measurable nondestructively, so that the same specimen can be monitored throughout the creep process.   This section uses the electrical resistivity as the property for nondestructive measurement.   Prior work that involved monitoring a material property during creep was limited to the stiffness [159] and the ultrasonic pulse velocity [160].

The creep resistance of cement-based materials is affected by admixtures such as silica fume [161-164], fly ash [161,163,165], slag [163] and steel fibers [166,167].   However, this section does not address the effect of admixtures.

Fig. 4.41 [168] shows the fractional change in electrical resistivity in the stress direction during creep testing at a constant compressive stress (20 MPa, compared to the compressive strength of 51 MPa) for plain cement mortar in the cured state (28 days of curing).   The fractional change in resistivity is essentially equal to the fractional change in resistance due to the small strain involved.

The resistivity increases as creep progresses except for the initial stage of creep, in which the resistivity drops slightly. The initial drop in resistivity is slight and does not occur in all the specimens. It is believed to be due to stress-induced healing of defects (Sec. 4.3.1.1).

The main effect of creep is the increase on resistivity, which is attributed to microstructural change, which can be viewed as minor damage. The effect of creep on the resistivity is consistent with that of strain rate on the resistivity: the lower the strain rate, the higher the fractional change in resistivity at the same strain (Fig. 4.17). The fractional change in resistivity per unit strain (Fig. 4.41, for the portion after the initial resistivity drop) is 500, which is larger than the value of 250 for static compressive strain (i.e., instantaneous strain rather than creep strain) (Fig. 4.17) [57]. This means that the extent of creep-induced microstructural change is larger than that of stress-induced microstructural change.

4.7 Conclusion

Electrical resistivity measurement provides a means of sensing the damage (or microstructural changes) in a cement-based material in real time. Damage is shown by an increase in the resistivity. For indicating the damage within the cement-based material, the volume resistivity is the relevant quantity. For indicating that at the interface between concrete rebar and concrete, between old concrete and new concrete, and between unbonded concrete elements, the contact resistivity is the relevant quantity. For indicating the damage at the interface between concrete and its carbon fiber epoxy-matrix composite retrofit, the apparent volume resistivity of the retrofit is the relevant quantity.

The fractional change in volume resistivity per unit strain is a parameter that describes the extent of strain-induced microstructural change in a cement-based material. This extent is thus found to be much larger for compressive creep and drying shrinkage than static compressive deformation. Creep occurs over time, thus allowing relatively extensive microstructural change to take place. Drying shrinkage is accompanied by the hydration reaction, which is necessarily accompanied by extensive microstructural change.

During static compression of mortar, the extent of microstructural change at a given strain, as indicated by the fractional change in resistivity, decreases with increasing loading rate. This is because it takes time for microstructural change to occur. The need for time is also indicated by the large effect observed during creep.

Freeze-thaw cycling causes damage, which progresses cycle by cycle and occurs in each cycle more significantly upon cooling than upon heating, as shown by the resistivity at the coldest point of a cycle increasing upon cycling more than that at the warmest point of a cycle.

Proximity to the edge of a piece of cast mortar prior to loading causes damage to a distance up to 50 mm from the edge, due to the faster drying at the edge than the center.

In contrast to damage, which causes the volume resistivity to increase, defect healing causes this resistivity to decrease. Defect healing occurs during compression, but subsequent unloading annuls the healing and aggravates the damage.

References

1. V. Pensee, D. Kondo and L. Dormieux, *J. Eng. Mech.* 128(8), 889-897 (2002).
2. F. Ragueneau and Ch. La Borderie, J. Mazars, *Mechanics of Cohesive-Frictional Materials* 5(8), 607-625 (2000).
3. R. Mahnken, D. Tikhomirov and E. Stein, *Computers & Structures* 75(2), 135-143 (2000).
4. L. Gao and C.-T.T. Hsu, *ACI Mater. J.* 95(5), 575-581 (1998).
5. A. Litewka, J. Bogucka and J. Debinski, *Archives of Civil Eng.* 42(4), 425-445 (1997).
6. S. Murakami and K. Kamiya, *Int. J. Mech. Sci.* 39(4), 473-486 (1997).
7. E. Papa and A. Taliercio, *Eng. Fracture Mech.* 55(2), 163-179 (1996).
8. D.J. Stevens and D. Liu, *J. Eng. Mech. – ASCE* 118(6), 1184-1200 (1992).
9. D. Breysse and N. Schmitt, *Cem. Concr. Res.* 21(6), 963-974 (1991).
10. S. Yazdani and H.L. Schreyer, *J. Eng. Mech. – ASCE* 116(7), 1435-1450 (1990).
11. W. Suaris, C. Ouyang and V.M. Fernando, *J. Eng. Mech. – ASCE* 116(5), 1020-1035 (1990).
12. D. Breysse, *Structural Safety* 8(1-4), 311-325 (1990).
13. H. Watanabe, Y. Murakami and M. Ohtsu, *Zairyo/J. Soc. Mater. Sci., Japan* 50(12), 1370-1374 (2001).
14. H. Ji, T. Zhang, M. Cai and Z. Zhang, *Yanshilxue Yu Gongcheng Xuebao/Chinese J. Rock Mechanics & Eng.* 19(2), 165-168 (2000).
15. E.N. Landis, *Construction & Building Materials* 13(1), 65-72 (1999).
16. T.J. Mackin and M.C. Roberts, *J. Amer. Ceramic Soc.* 83(2), 337-343 (2000).
17. Y. Chen, N. Li, A. Li, Y. Pu and Q. Liao, *Yanshilixue Yu Gongcheng Xuebao/Chinese J. Rock Mechanics & Eng.* 19(6), 702-706 (2000).
18. H.I. Schreyer and M.K. Neilsen, *ASCE Eng. Mech. Specialty Conf. on Mechanics Computing in 1990's and Beyond,* ASCE, New York, NY, 303-307 (1991).

19. H. Cordes and D. Bick, *Beton – und Stahlbetonbau* 86(8), 181 (1991).
20. H. Meichsner, *Beton – und Stahlbetonbau* 87(4), 95 (1992).
21. A. Mor, P.J.M. Monteiro and W.T. Hester, *Cem., Concr. & Aggregates* 11(2) 121 (1989).
22. C. Edvardsen, *Betonwerk und Fertigteil-Technik* 62(11), 77 (1996).
23. S. Jacobsen, J. Marchand and L. Boisvert, *Cem. Concr. Res.* 26(6), 869 (1996).
24. S. Jacobsen and E.J. Sellevold, *Cem. Concr. Res.* 26(1), 55 (1996).
25. S. Jacobsen, J. Marchand and H. Hornain, *Cem. Concr. Res.* 25(8), 1781 (1995).
26. N. Hearn, *Mater. & Struct.* 31(212), 563 (1998).
27. C.M. Dry, *Proc. 3rd Int. Conf. on Intelligent Materials and 3rd European Conf. on Smart Struct. and Mater.*, SPIE – the International Society for Optical Engineering, Society of Photo-Optical Instrumentation Engineers, Bellingham, WA, 2779, 958 (1996).
28. C.M. Dry, *Proc. 3rd Int. Conf. on Intelligent Mater. and 3rd European Conf. on Smart Struct. and Mater.*, SPIE – the International Society for Optical Engineering, Society of Photo-Optical Instrumentation Engineers, Bellingham, WA, 2719, 247 (1996).
29. C. Dry, *Smart Mater. & Struct.* 3(2), 118 (1994).
30. M.P. Luong, *Mechanics of Deformation and Flow of Particulate Materials*, ASME-ASCE-SES Joint Summer Meeting, ASCE, New York, NY, 199-213 (1997).
31. O. Buyukozturk and H.C. Rhim, *Construction & Building Mater.* 11(3), 195-198 (1997).
32. P. Mallinson, *Insight-non-Destructive Testing & Condition Monitoring* 39(12), 874-877 (1997).
33. S.W. Hearn and C.K. Shield, *ACI Mater. J.* 94(6), 510-519 (1997).
34. D. Tarchi, E. Ohlmer and E. Grinzato, *Research in Nondestructive Evaluation* 9(4), 181-200 (1997).
35. V. Vavilov, T. Kauppinen and E. Grinzato, *Research in Nondestructive Evaluation* 9(4), 181-200 (1997).
36. M. Sansalone, *ACI Structural J.* 94(6), 777-786 (1997).
37. W. Uddin and F. Najafi, *Proc. Speciality Conf. Infrastructure Condition Assessment: Art, Science, Practice*, ASCE, New York, NY, 524-533 (1997).
38. D.J. Transue, K.L. Rens and M.P. Schuller, *Proc. Speciality Conf. Infrastructure Condition Assessment: Art, Science, Practice*, ASCE, New York, NY, 415-424 (1997).
39. S.K. Park and T. Uomoto, *Insight-non-Destructive Testing & Condition Monitoring* 39(7), 488-493 (1997).
40. Y. Wang and D.D.L. Chung, *Cem. Concr. Res.* 28(10), 1373-1378 (1998).

41.   X. Fu and D.D.L. Chung, *Cem. Concr. Res.* 25(4), 689-694 (1995).
42.   J. Cao, S. Wen and D.D.L. Chung, *J. Mater. Sci.* 36(18), 4351-4360 (2001).
43.   Z. Bayasi and J. Zhou, *ACI. Mater. J.* 90(4), 349 (1993).
44.   B. Ma, J. Li and J. Peng, *J. Wuhan University of Technology*, Materials Science Edition 14(2), 1 (1999).
45.   K. Tan and X. Pu, *Cem. Concr. Res.* 28(12), 1819 (1998).
46.   L. Bagel, *Cem. Concr. Res.* 28(7), 1011 (1998).
47.   C.A. Ross, *Proc. 1997 ASME Pressure Vessels and Piping Conf. on Structures Under Extreme Loading Conditions*, American Society of Mechanical Engineers, Pressure Vessels and Piping Division, New York, NY, 351, 255-262 (1997).
48.   R. John and S.P. Surendra, *Fracture of Concrete and Rock: SEM-RILEM Int. Conf.*, Society for Experimental Mechanics Inc., Bethel, CT, 35-52 (1987).
49.   S. Ohgishi and H. Ono, *Zairyo/J. Soc. Mater. Sci.* 29(318, 279-285 (1980).
50.   Z. Li and Y. Huang, *ACI Mater. J.*, 95(5), 512-518 (1998).
51.   S. Mindess, *Application of Fracture Mechanics to Cementitious Composites, NATO ASI Series, Series E: Applied Sciences*, Martinus Nijhoff Publ., Dordrecht, Netherlands and Boston, MA, (94), 617-636 (1984).
52.   J.-I. Takeda, *Cement-Based Composites: Strain Rate Effects on Fracture, Mater. Res. Soc. Symp. Proc.*, Materials Research Society, Pittsburgh, PA, 64, 15-20 (1985).
53.   D. Chandra, *Proc. 10th Conf. Eng. Mechanics*, ASCE, New York, NY, 1, 102-105 (1995).
54.   D. Chandra and T. Krauthammer, *Earthquake Eng. & Struct. Dynamics*, 24(12), 1609-1622 (1995).
55.   J.-H. Yon, N.M. Hawkins and A.S. Kobayashi, *ACI Mater. J.* 89(2), 146-153 (1992).
56.   E. Pozzo, *Mater. & Struct.* 20(118), 303-314 (1987).
57.   J. Cao and D.D.L. Chung, *Cem. Concr. Res.* 32(5), 817-819 (2002).
58.   P. Chen and D.D.L. Chung, *Composites: Part B* 27B, 11-23 (1996).
59.   X. Fu and D.D.L. Chung, *Cem. Concr. Res.* 26(1), 15-20 (1996).
60.   D. Bontea, D.D.L. Chung and G.C. Lee, *Cem. Concr. Res.* 30(4), 651-659 (2000).
61.   F. Reza, G.B. Batson, J.A. Yamamuro and J.S. Lee, *ACI Publication SP 206*, Concrete: Materials Science to Applications: A Tribute to Surendra P. Shah, American Concrete Institute, Farmington Hills, MI, p. 429-438 (2002).
62.   Z. Li, M. Xu and N.C. Chung, *Mag. Concr. Res.* 50(1), 49-57 (1998).
63.   V.A. Ghio and P.J.M. Monteiro, *ACI Mater. J.* 94(2), 111-118 (1997).

64. C.K. Kankam, *J. Struct. Eng. – ASCE* 123(1), 79-85 (1997).

65. H.P. Schroeder and T.B. Wood, *J. Cold Regions Eng.* 10(2), 93-117 (1996).

66. N.M. Ihekwaba, B.B. Hope and C.M. Hansson, *Cem. Concr. Res.* 26(2), 267-282 (1996).

67. A.A. Almusallam, A.S. Al-Gahtani and A.R. Aziz, *Construction & Building Mater.* 10(2), 123-129 (1996).

68. B.S. Hamad, *ACI Mater. J.* 92(6), 579-590 (1995).

69. A. Hamouine and M. Lorrain, *Mater. Struct.* 28(184), 569-574 (1995).

70. K. Thangavel, N.S. Rengaswamy and K. Balakrishnan, *Indian Concr. J.* 69(5), 289-293 (1995).

71. B.S. Hamad, *ACI Struct. J.* 92(1), 3-13 (1995).

72. F. de Larrard, I. Schaller and J. Fuchs, *ACI Mater. J.* 90(4), 333-339 (1993).

73. A.R. Cusens and Z. Yu, *Cem. Concr. Compos.* 14(4), 269-276 (1992).

74. A.R. Cusens and Z. Yu, *Struct. Eng.* 71(7), 117-124 (1993).

75. A. Mor, *ACI Mater. J.* 89(1), 76-82 (1992).

76. M. Maslehuddin, I.M. Allam, G.J. Al-Sulaimani, A. Al-Mana and S.N. Abduljauwad, *ACI Mater. J.* 87(5), 496-502 (1990).

77. C.-H. Chiang, C.-L. Tsai and Y.-C. Kan, *Ultrasonics* 38(1), 534-536 (2000).

78. G.L. Balazs, C.U. Grosse, R. Koch and H.W. Reinhardt, *Mag. Concr. Res.* 48(177), 311-320 (1996).

79. C.-H. Chiang and C.-K. Tang, *Ultrasonics* 37(3), 223-229 (1999).

80. X. Fu and D.D.L. Chung, *ACI Mater. J.* 95(6), 725-734 (1998).

81. J. Cao and D.D.L. Chung, *Cem. Concr. Res.* 31(4), 669-671 (2001).

82. G.K. Ray, *Concr. Int.:* Design & Construction 9(6), 24-28 (1987).

83. E.K. Schrader, *Concr. Int.:* Design & Construction 14(11), 54-59 (1992).

84. C. Ozyildirim, *ACI SP-132*, V.M. Malhotra (Ed.), American Concrete Institute, Detroit, 1992, p. 1287.

85. M.D. Luther, *Transp. Res. Rec.* (1204), 11-20 (1988).

86. D.G. Manning and J. Ryell, *Transp. Res. Rec.* (762), 1-9 (1980).

87. L. Calvo, M. Meyers, *Concr. Int.:* Design & Construction 13(7), 46-47 (1991).

88. J.K. Bhargava, *ACI SP-69*, D.W. Fowler, L.E. Kukacka (Eds.), American Concrete Institute, Detroit, 1981, p. 205.

89. P.-W. Chen, X. Fu and D.D.L. Chung, *Cem. Concr. Res.* 25(3) 491-496 (1995).

90. J. Cao and D.D.L. Chung, *Cem. Concr. Res.* 31(11), 1647-1651 (2001).

91. X. Luo and D.D.L. Chung, *Cem. Concr. Res.* 30(2), 323-326 (2000).

92. H.A. Toutanji, *Compos. Struct.* 44(2), 155-161 (1999).

93.   H.A. Toutanji and T. El-Korchi, *J. Compos. Construction* 3(1), 38-45
       (1999).
94.   Y. Lee and S. Matsui, *Tech. Reports Osaka University* 48(2319-2337),
       247-254 (1998).
95.   E.K. Lau, A.S. Mosallam and P.R. Chakrabarti, *Int. SAMPE Tech.
       Conf.*, SAMPE, Covina, CA, 30, 293-302 (1998).
96.   A. Liman and P. Hamelin, *Proc. Int. Conf. Computer Methods in
       Composite Materials*, CADCOMP, Computational Mechanics Publ.,
       Ashurst, England, 569-578 (1998).
97.   P. Balaguru and S. Kurtz, *Proc. Int. Seminar on Repair and
       Rehabilitation of Reinforced Concrete Structures: The State of the Art*,
       ASCE, Reston, VA, 155-168 (1998).
98.   A. Nanni and W. Gold, *Proc. Int. Seminar on Repair and
       Rehabilitation of Reinforced Concrete Structures: The State of the Art*,
       ASCE, Reston, VA, 144-154 (1998).
99.   H.A. Toutanji and T. El-Korchi, *Proc. Int. Seminar on Repair and
       Rehabilitation of Reinforced Concrete Structures: The State of the Art*,
       ASCE, Reston, VA, 134-143 (1998).
100.   Z. Geng, M.J. Chajes, T. Chou and D.Y. Pan, *Compos. Sci. Tech.*
       58(8), 1297-1305 (1998).
101.   I. Gergely, C.P. Pantelides, R.J. Nuismer and L.D. Reaveley, *J.
       Compos. Construction* 2(4), 165-174 (1998).
102.   H.N. Garden and L.C. Hollaway, *Compos., Part B* 29(4), 411-424
       (1998).
103.   K. Kikukawa, K. Mutoh, H. Ohya, Y. Ohyama, H. Tanaka and K.
       Watanabe, *Compos. Interfaces* 5(5), 469-478 (1998).
104.   M. Xie and V.M. Karbhari, *J. Compos. Mater.* 32(21), 1894-1913
       (1998).
105.   V.M. Karbhari, M. Engineer and D.A. Eckel II, *J. Mater. Sci.* 32(1),
       147-156 (1997).
106.   D.P. Henkel and J.D. Wood, *NDT & e International* 24(5), 259-264
       (1991).
107.   A.K. Pandey and M. Biswas, *J. Sound & Vibration* 169(1), 3-17
       (1994).
108.   Z. Mei and D.D.L. Chung, *Cem. Concr. Res.* 30(5), 799-802 (2000).
109.   H. Marzouk and D. Jiang, *ACI Mater. J.*, 91(6), 577-586 (1994).
110.   L. Biolzi, G.L. Guerrini and G. Rosati, *J. Mater. Civil Eng.* 11(2), 167-
       170 (1999).
111.   J.H. Rutherford, B.W. Langan and M.A. Ward, *Cem., Concr. &
       Aggregates* 16(1), 78-82 (1994).
112.   D. Bordeleau, M. Pigeon and N. Banthia, *ACI Mater. J.* 89(6), 547-553
       (1992).
113.   B.B. Sabir, *Cem. & Concr. Compos.* 19(4), 285-294 (1997).

114. H. Mori, Y. Ishikawa, T. Shibata and T. Okamoto, *Zairyo/J. Soc. Mater. Sci.* Japan 48(8), 889-894 (1999).

115. T. Bakharev and L.J. Struble, *Proc. 1994 MRS Fall Meeting on Microstructure of Cement-Based Systems/Bonding and Interfaces in Cementitious Materials,* Materials Research Society, Pittsburgh, PA, 370, 83-88 (1995).

116. N.M. Akhras, *Cem. Concr. Res.* 28(9), 1275-1280 (1998).

117. H. Cai and X. Liu, *Cem. Concr. Res.* 28(9), 1281-1287 (1998).

118. S. Wen and D.D.L. Chung, *Cem. Concr. Res.* 29(6), 961-965 (1999).

119. J. Cao and D.D.L. Chung, *Cem. Concr. Res.* 32(10), 1657-1661 (2002).

120. V. Baroghel-Bouny and J. Godin, *Bulletin de Liaison des Laboratoires des Ponts et Chaussees,* (218), 39-48 (1998).

121. M.D. Luther and P.A. Smith, *Proc. Eng. Foundation Conf.,* 75-106 (1991).

122. V.M. Malhotra, *Concr. Int.: Design & Construction* 15(4), 23-28 (1993).

123. M.D. Luther, *Concr. Int.: Design & Construction* 15(4), 29-33 (1993).

124. J. Wolsiefer and D.R. Morgan, *Concr. Int.: Design & Construction* 15(4), 34-39 (1993).

125. D.D.L. Chung, *J. Mater. Sci.* 27(8), 673-682 (2002).

126. Y. Xu and D.D.L. Chung, *Cem. Concr. Res.* 30(8), 1305-1311 (2000).

127. M.N. Haque, *Cem. Concr. Compos.* 18(5), 333-342 (1996).

128. S.H. Alsayed, *Cem. Concr. Res.* 28(10), 1405-1415 (1998).

129. A. Lamontagne, M. Pigeon, R. Pleau and D. Beaupre, *ACI Mater. J.* 93(1), 69-74 (1996).

130. M.G. Alexander, *Adv. Cem. Res.* 6(22), 73-81 (1994).

131. F.H. Al-Sugair, *Mag. Concr. Res.* 47(170), 77-81 (1995).

132. B. Bissonnette and M. Pigeon, *Cem. Concr. Res.* 25(5), 1075-1085 (1995).

133. G.A. Rao, *Cem. Concr. Res.* 28(10), 1505-1509 (1998).

134. Z. Li, M. Qi, Z. Li and B. Ma, *J. Mater. Civil Eng.* 11(3), 214-223 (1999).

135. O.M. Jensen and P.F. Hansen, *ACI Mater. J.* 93(6), 539-543 (1996).

136. O.M. Jensen, P.F. Hansen, *Adv. Cem. Res.* 7(25), 33-38 (1995).

137. P.-W. Chen and D.D.L. Chung, *Composites* 24(1), 33-52 (1993).

138. J. Cao and D.D.L. Chung, *Cem. Concr. Res.,* in press.

139. S.K. Kaushik, V.P. Bhargava and V. Kumar, *Indian Concr. J.* 75(8), 515-521 (2001).

140. J. Li and Y. Yao, *Cem. Concr. Res.* 31(8), 1203-1206 (2001).

141. V. Sicard, R. Francois, E. Ringot and G. Pons, *Cem. Concr. Res.* 22(1), 159-168 (1992).

142. Y. Ishikawa, H. Kikukawa and T. Tanabe, *Transactions of the Japan Concrete Institute* 18, 107-114 (1996).

143.   G. De Schutter and L. Taerwe, *Mater. Struct.* 33(230), 370-380 (2000).
144.   K. Kovler, *J. Mater. Civil Eng.* 11(1), 84-87 (1999).
145.   G. De Schutter and L. Taerwe, *Mag. Concr. Res.* 49(180), 195-200 (1997).
146.   A.A. Khan, W.D. Cook and D. Mitchell, *ACI Mater. J.* 94(2), 156-163 (1997).
147.   Z. Li, *Int. J. Fracture* 66(2), 189-196 (1994).
148.   A.S. Ngab, A.H. Nilson and F.O. Slate, *J. Amer. Concr. Inst.* 78(4), 255-261 (1981).
149.   M.M. Smadi and F.O. Slate, *ACI Mater. J.* 86(2), 117-127 (1989).
150.   Z.P. Bazant and J.-K. Kim, *Mater. & Struct.* 24(144), 409-421 (1991).
151.   R.N. Swamy and G.H. Lambert, *Proc. – 2nd Int. Conf., Publication SP – American Concrete Institute 91*, American Concrete Institute, Detroit, MI, 1, 145-170 (1991).
152.   F. Furumura, T. Ave and W.J. Kim, *Report of the Research Laboratory of Engineering Materials, Tokyo Institute of Technology* (11), 183-199 (1986).
153.   S. Ohgishi, M. Wada and H. Ono, *Proc. – Computer Networking Symposium,* Pergamon Press, Oxford, Engl. and Elmsford, NY, 3, 109-119 (1980).
154.   S.G. Reid and C. Qin, *2nd National Structural Eng. Conf. 1990,* National Conference Publication – Institution of Engineers, IE Aust, Barton, Australia, (90), pt. 10, 255-258 (1990).
155.   H. Gao, G. Liu and F. Chen, *Qinghua Daxue Xuebao/J. Tsinghua University* 41(11), 110-113 (2001).
156.   E.A. Kogan, *Hydrotechnical Construction (Gidrotekhnicheskoe Stroitel'Stvo)* 17(9), 448-452 (1983).
157.   J.L. Clement and F. Le Maou, *Bull. de Liaison des Laboratoires des Ponts et Chaussees* (228), 333-339 (1993).
158.   A. Benaissa, P. Morlier and C. Viguier, *Mater. & Struct.* 26(160), 333-339 (1993).
159.   E.K. Attiogbe and D. Darwin, *ACI Mater. J.* 85(1), 3-11 (1988).
160.   J. Zhu, L. Chen and X. Yan, *Gong Cheng Li Xue/Engineering Mechanics* 15(3), 111-117 (1998).
161.   S. Ghosh and K.W. Nasser, *Canadian J. Civil Eng.* 22(3), 621-636 (1995).
162.   K. Wiegrink, S. Marikunte and S.P. Shah, *ACI Mater. J.* 93(5), 409-415 (1996).
163.   R.P. Khatri, V. Sirivivatnanon and W. Gross, *Cem. Concr. Res.* 25(1), 209-220 (1995).
164.   W.A. Al-Khaja, *Construction & Building Materials* 8(3), 169-172 (1994).
165.   R.S. Ghosh and J. Timusk, *J. Amer. Concr. Inst.* 78(5), 351-357 (1981).

166.   P.S. Mangat and A.M. Motamedi, *Mater. & Struct.* (113), 361-370 (1986).

167.   P.S. Mangat and M.M. Azari, *J. Mater. Sci.* 20(3), 1119-1133 (1985).

168.   J. Cao and D.D.L. Chung, *Cem. Concr. Res.*, in press.

5

Electrically Conductive Cement-Based Materials

5.1 Introduction

Cement-based materials have received much attention in relation to their mechanical properties, due to their importance as structural materials. However, the need for a structural material to be able to serve one or more non-structural functions while retaining good structural properties is increasingly recognized [1]. This is because the use of a multifunctional structural material in place of a combination of a structural material and a nonstructural functional material (e.g., a structural material with an embedded nonstructural functional material) reduces cost, enhances durability and repairability, increases the functional volume, avoids degradation of the mechanical properties, and simplifies design. Nonstructural functions include sensing, actuation, heating, corrosion protection, self-healing, thermal insulation, heat retention and electromagnetic interference (EMI) shielding [1,2].

One category of multifunctional cement-based materials is electrically conductive cement-based materials [3,4]. The conductivity is attractive for electrical grounding, lightning protection, resistance heating (e.g., in deicing and building heating), static charge dissipation, electromagnetic interference shielding, thermoelectric energy generation and for overlays (electrical contacts) used in the cathodic protection of steel reinforcing bars (rebars) in concrete.

5.2   Electrical conduction

The cement matrix is electrically attractive due to its electrical conductivity, which is in contrast to the nonconductive behavior of most polymers. Due to the conductivity of the cement-matrix, an electrically conductive admixture (i.e., a conductive filler) in a cement-matrix composite can enhance the conductivity of the composite even when the volume fraction of the admixture is below the percolation threshold, which refers to the volume fraction above which the admixture units touch to form a continuous conduction path. The percolation threshold is determined from the variation of the electrical resistivity with the volume fraction of the conductive admixture. The electrical resistivity abruptly decreases by orders of magnitude at the percolation threshold (Fig. 2.28) [3]. In most cases, the percolation threshold decreases with increasing aspect ratio and with decreasing unit size of the admixture. In the case of short carbon fibers (7 μm diameter) in cement, the percolation threshold decreases with increasing fiber length from 1 to 10 mm [5]. However, the percolation threshold also depends on the unit size of the nonconductive or less conductive components in the composite. Thus, the presence of sand (a nonconductive component) affects the percolation threshold (Fig. 2.28) [2]. In the absence of sand, the percolation threshold is between 0.5 and 1.0 vol.% when the conductive admixture in cement is short carbon fiber (15 μm diameter, 5 mm long) (Fig. 2.28) [3].

The curing age has relatively minor influence on the electrical resistivity, although it has major influence on the mechanical properties [6], as described in Sec. 2.4. Nevertheless, the effect in the absence of conducting fibers, especially in terms of the impedance, is sufficient for use in studying the curing process [7-10]. An increase in the carbon fiber content from 0.53 to 1.1 vol. % diminishes the effect of curing age significantly, because the fibers become more dominant in governing the resistivity as the fiber content increases.

Cement paste is electrically conductive, with DC resistivity at 28 days of curing around 5×10^5 Ω.cm at room temperature. The resistivity is increased slightly (to 6×10^5 Ω.cm) by the addition of silica fume (SiO_2 particles around 0.1 μm in size, in the amount of 15% by mass of cement), and is increased more (to 7×10^5 Ω.cm) by addition of latex (20% by mass of cement), which is a styrene-butadiene copolymer in the form of particles of size around 0.2 μm [11]. The higher the latex content, the higher is the resistivity [12]. In case of mortars (with fine aggregate, i.e., sand), the transition zone between the cement paste and the aggregate enhances the conductivity [13]. Whether aggregates (sand and stones) are present or not, the AC impedance spectroscopy technique for characterizing the frequency-dependent electrical behavior is useful for studying the microstructure [13-16].

The nonconductive admixture effects on the resistivity, as mentioned above, are small compared to the effect of adding short conductive fibers. Nevertheless, the nonconductive admixtures can help the fiber dispersion, thereby causing the resistivity of cement-based materials containing conductive short fibers to be lower. At a volume fraction below the percolation threshold, the electrical conductivity of a composite is highly dependent on the degree of fiber dispersion. The greater the degree of fiber dispersion, the higher the conductivity of the composite. This is because of the relatively long length of the conduction path within the matrix in case of poor fiber dispersion, as illustrated in Fig. 5.1. At the same carbon fiber (15 μm diameter) volume fraction (0.35 vol. %, below the percolation threshold), the resistivity of cement mortar is lower when silica fume is present along with the fibers, due to the effectiveness of silica fume in helping the fiber dispersion [17]; it is further

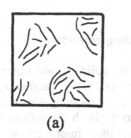 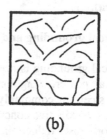

(a) (b)

Fig. 5.1 Fiber dispersion below the percolation threshold. (a) Poor dispersion. (b) Good dispersion. From Ref. 19.

lowered when both methylcellulose and silica fume are present along with the fibers [17,18]. The use of acrylic, styrene acrylic or latex dispersions in place of the methylcellulose solution is less effective for lowering the resistivity in the presence of the carbon fibers, but it results in higher values of the tensile ductility. The acrylic dispersion is more effective than styrene acrylic or latex dispersions in enhancing the tensile ductility and the tensile strength [17]. At the same steel fiber (60 μm diameter) volume fraction (0.05 vol.%, much below the percolation threshold), the resistivity of cement mortar is lower when silane is present along with the fibers, due to the effectiveness of silane in helping the fiber dispersion [19].

The electrical conductivity of a cement-based material containing a conductive admixture is governed by the conductivity of the admixture itself, the degree of dispersion of the admixture and the contact electrical resistivity of the interface between the admixture and the cement matrix. Due to the conductivity of the cement matrix, this contact resistivity is important, particularly when the admixture volume fraction is below the percolation threshold. The contact electrical resistivity between stainless steel fiber (60 μm diameter) and cement paste is around 6×10^6 Ω.cm² and is smaller if the fiber has been acid washed [20].

The interface between steel fiber and the cement matrix behaves similarly to that between steel rebar and concrete. The latter is more common in practice than the former. The contact resistivity of the latter interface is around 6×10^7 Ω.cm² [21].

5.3   Applications

5.3.1   Electrical grounding and lighting protection

Electrical grounding is needed for buildings and other structures which involve electrical power. Lightning protection is needed for tall buildings. Metals such as steel are commonly used for these applications. However, the use of electrically conductive concrete to diminish the volume of metal required is attractive for cost reduction, durability improvement and installation simplification.

5.3.2   Static charge dissipation

Static charge dissipation is needed for structures that come into contact with sensitive electronic devices. Metals and conductor-filled polymer-matrix composites are used for this purpose. However, the use of electrically conductive concrete for this application allows large volumes of structure to have the ability for static charge dissipation.

5.3.3   Resistance heating

Due to the environmental problem associated with the use of fossil fuels and due to the high cost of solar heating, electrical heating is increasingly important. Although electric heat pumps are widely used for the electrical heating of buildings, resistance heating is a complementary method which is receiving increasing attention due to the low costs for its implementation and control, its adaptability to localized heating (e.g., the heating of a particular room of a building), its nearly 100% efficiency of conversion of electrical energy to heat energy, and the increasing demand of safety and the quality of life. Resistance heating is needed in buildings and for the deicing of driveways, bridges, highways and airport runways. Deicing is valuable for hazard mitigation. The alternate technique of snow removal (shoveling) is labor intensive and takes time, in contrast to the automatic and continuous nature of deicing by resistance heating.

Resistance heating involves passing an electric current through a resistor, which is the heating element. In relation to the heating of buildings and other structures, resistance heating typically involves the embedding of heating elements in the structural material, such as concrete. The materials of heating elements cannot be too low in electrical resistivity, as this would result in the resistance of the heating element being too low. The materials of heating elements cannot be too high in resistivity either, as this would result in the current in the heating element being too low (unless the voltage is very high). Materials of heating elements are commonly metal alloys such as nichrome.

Thus metal wires are commonly embedded in a structural material in order to provide resistance heating. However, the embedding degrades the mechanical properties of the structural component, and the repair of the embedded heating element is difficult. Furthermore, the embedding is limited to selected locations of a structural component, and consequently the heating is not uniform. The nonuniformity is worsened by the poor thermal conductivity of the structural materials. An electrically conductive cement-based material can be used as a resistance (Joule) heating element [22,23]. There is no need to embed wires in the structural component, thereby alleviating the problems mentioned above in connection with the embedment.

Conventional concrete is electrically conductive, but the resistivity is too high for efficient heating purposes. The resistivity of concrete can be diminished by the introduction of electrically conductive material(s) into concrete, such as discontinuous fibers and particles. A comparative evaluation of the effectiveness of discontinuous carbon fibers (15 μm diameter, 1.0 vol.%), discontinuous stainless steel fibers (8 μm diameter, 0.7 vol.%) and

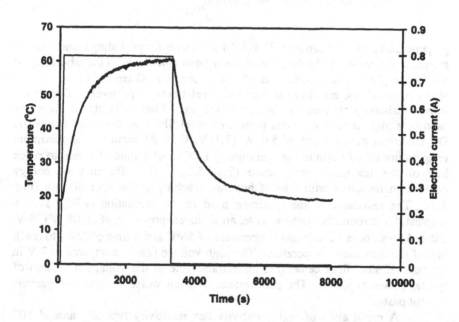

Fig. 5.2   Temperature variation during heating (current on) and subsequent cooling (current off), using steel fiber cement as the resistance heating element. Thick curve: temperature. Thin curve: current.

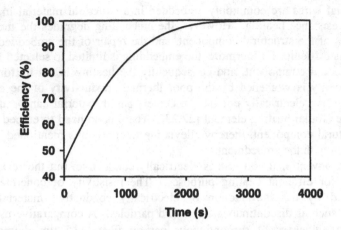

Fig. 5.3   Temperature variation during heating (current on) and subsequent cooling (current off). Thick curve: temperature. Thin curve: current.

graphite particles (<45 μm size, 37 vol.%) in cement for providing cement-based materials for resistance heating has recently been completed in the laboratory of the author [22]. The resistivity is 104, 407 and 0.85 Ω.cm for cement pastes with carbon fibers, graphite particles and steel fibers respectively. Due to the lowest resistivity attained by the use of the steel fibers, the effectiveness for heating is highest for the cement paste with steel fibers, as described below. A DC electrical power input of 5.6 W (7.1 V, 0.79 A) results in a maximum temperature of 60°C (initial temperature = 19°C) and a time of 6 min to reach half of the maximum temperature (Fig. 5.2). The efficiency of energy conversion increases with time of heating, reaching 100% after 50 min (Fig. 5.3). The resistance of the specimen used in the evaluation is 8.80 Ω. In contrast, for carbon fiber cement paste, an electrical power input of 1.8 W (28 V, 0.065 A) results in a maximum temperature of 56°C and a time of 256 s to reach half of the maximum temperature. The high voltage (28 V, compared to 7 V in the case of steel fiber cement) is undesirable due to the voltage limitation of typical power supplies. The performance is even worse for graphite particle cement paste.

A metal alloy of high resistivity has resistivity typically around 10^{-4} Ω.cm. For a metal wire of diameter 1 mm, a length of 7 m is needed to attain a resistance of 8.8 Ω, which is the value attained in the abovementioned steel fiber cement of size 150 x 12 x 12 mm (i.e., a length of only 0.15 m). Thus, the use of a cement-based heating element does not require a long length. In contrast, a metal-based heating element requires a long length, and consequently winding of the long length to make a heater coil is necessary. Due to its bulkiness, a metal coil is intrusive when it is embedded in a structure. Furthermore the air gap within the coil is a thermal insulator that reduces the effectiveness of heat transfer from the coil to the structure.

5.3.4 Cathodic protection

Cathodic protection is one of the most common and effective methods for corrosion control of steel-reinforced concrete [24-28]. This method involves the application of a voltage so as to force electrons to go to the steel rebar, thereby making the steel a cathode. For directing electrons to the steel-reinforced concrete to be cathodically protected, an electrical contact is needed on the concrete. The electrical contact is electrically connected to the voltage supply. One of the choices of an electrical contact material is zinc, which is a coating deposited on the concrete by thermal spraying. It has a very low volume resistivity (thus requiring no metal mesh embedment), and it can serve both as a sacrificial anode and as an electrical contact, but it suffers from poor wear and corrosion resistance, the tendency to oxidize, high thermal expansion coefficient, and high material and processing costs. Another choice is a conductor filled polymer [29], which can be applied as a coating without heating and can be used alone or as an adhesive between concrete and a zinc plate, but it suffers from poor wear resistance, high thermal expansion coefficient and high material cost. Yet another choice is a metal (e.g., titanium) strip or wire embedded at one end in cement mortar, which is in the form of a coating on the steel reinforced concrete. The use of electrically conductive mortar for this coating facilitates cathodic protection by reducing the required running voltage, thereby saving energy [30].

5.3.5 Electromagnetic interference shielding

Electrically conductive cement-based materials are also attractive for EMI shielding [31-35] (Sec. 2.9), which is in demand due to the interference of wireless (particularly radio frequency, i.e., frequency $10^4 - 10^{10}$ Hz, wavelength $10^{-1} - 10^4$ m, and photon energy $10^{-10} - 10^{-5}$ eV) devices with digital devices and the increasing sensitivity and importance of electronic devices. Shielding is particularly needed for underground vaults containing transformers and other electronics that are relevant to electric power and telecommunication. It is also needed for deterring electromagnetic forms of spying. An EMI shield is a barrier to the transmission of electromagnetic fields in both directions. The EMI shielding effectiveness is the ratio of the magnitude (amplitude) of the field incident on the barrier to that transmitted through the barrier. It is equivalent to the ratio of the field incident on the electronics with the shield removed to that with the shield in place. Although a material that is effective for EMI shielding tends to be electrically conductive, a material that is superior in conductivity is not necessarily also superior in EMI shielding, as will be shown in Sec. 5.5.

The main mechanism for EMI shielding using conductive materials (e.g., carbon and steel) is reflection [35]. The loss (attenuation) due to reflection increases with decreasing frequency. However, another mechanism is absorption, as enhanced by electric and magnetic dipoles in the material. An example of a component with magnetic dipoles is Fe_2O_3, which is present in fly ash (a byproduct of coal combustion that is sometimes used as an admixture or as a cement replacement to decrease cost and to improve the resistance to alkali-silica reaction, sulfate attack and corrosion of steel reinforcement). The loss due to absorption increases with increasing frequency. The third mechanism, multiple reflections off the external surfaces and internal surfaces and interfaces of the material, is only important when the specimen is very thin or when the specimen has a great deal of internal surfaces or interfaces.

5.3.6 Lateral guidance in automatic highways

The ability to reflect electromagnetic radiation (particularly radio waves) is useful for automatic lateral guidance of vehicles. Lateral guidance as attained by steering is limited in safety due to the tendency for human error. It is also limited in accuracy, as shown by the difficulty of steering a car through a narrow lane or parking a car very close to a curb.

Automatic lateral guidance is needed for automatic highways [36], which refer to highways that provide fully automated control of vehicles, so that safety and mobility are enhanced. In other words, a driver does not need to drive on an automatic highway, as the vehicle goes automatically, with both lateral control (steering to control position relative to the center of the traffic lane) and longitudinal control (speed and headway).

Instead of human steering, lateral guidance can involve electromagnetic or magnetic interaction between a car and a lane, so that the reliance on human steering is reduced or removed, thereby enhancing safety and mobility, and facilitating parking a car very close to a curb (as needed by buses and by electric vehicles that require battery recharging).

This alternate form of lateral guidance currently involves the use of magnetic sensors in the cars together with magnetic highway marking. When a car deviates from its path, which is marked by magnets embedded along the length of the pavement, the magnetic sensor in the car detects the deviation. The signal from the sensor is then used to control the steering automatically in real time. In contrast, this chapter uses an electromagnetic form of lateral guidance, as made possible by radio wave-reflecting concrete.

Radio wave-reflecting concrete is concrete which contains an electrically conductive admixture, which causes the concrete to be a strong reflector of radio waves. Conventional concrete is a poor reflector. By coating

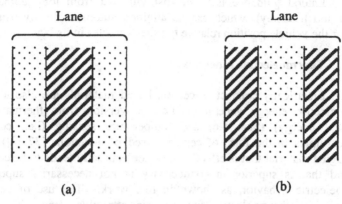

(a)   (b)

Fig. 5.4   Radio wave-reflecting concrete (or mortar), as indicated
                    by the shaded portion(s) of a lane and conventional
                    concrete, as indicated by the dotted portion(s) of a lane,
                    for attaining electromagnetic lateral guidance of vehicles.
                    (a) has the radio wave-reflecting concrete (or mortar) in
                    the middle part of lane, but (b) has the radio wave-
                    reflecting concrete (or mortar) in the edge portions of the
                    lane.

either the middle portion (Fig. 5.4(a)) or the edge portions (Fig. 5.4(b)) of a lane
of a highway with radio wave-reflecting concrete (or mortar) and by installing in
each vehicle a transmitter and a detector of radio wave, a vehicle can sense its
lateral position relative to the middle portion of the lane through the intensity of
the radio wave bounced back by the pavement.

Compared to the magnetic technology, the attractions of the
electromagnetic technology are low material cost (reflecting concrete, though
more expensive than conventional concrete, is much less expensive than
concrete with embedded magnets or magnetic strips), low labor cost (same as
conventional concrete, thus much less than concrete with embedded magnets or
magnetic strips), low peripheral electronic cost (off-the-shelf oscillator and
detector), good mechanical properties (reflecting concrete exhibits better
mechanical properties and lower drying shrinkage than conventional concrete,
whereas embedded magnets weaken concrete), good reliability (less affected by
weather, as frequency, impedance and power selectivity provides tuning
capability), and high durability (demagnetization and marking detachment not
being issues). Moreover, the magnetic field from a magnetic marking can be
shielded by electrical conductors (such as steel) between the marking and the
vehicle, whereas the electromagnetic field cannot be easily shielded.

Lateral guidance is to be distinguished from longitudinal guidance (speed and headway), which can be attained automatically by using radar to monitor the vehicle position relative to other vehicles in its lane.

5.3.7   Thermoelectric functions

Electrically conductive cement-based materials are also attractive for thermoelectric energy generation (i.e., conversion from thermal energy to electrical energy) (Sec. 5.6) and temperature sensing (i.e., cement-based thermocouples in the form of cement-based pn junctions [37,38]) (Sec. 2.8). Although a material that is effective thermoelectrically tends to be conductive, a material that is superior in conductivity is not necessarily superior in the thermoelectric behavior, as shown in this work. The use of cement-based materials for thermoelectric functions is attractive, since this allows the functions to be built into concrete structures [39-43]. Due to the large volume of concrete structures and the low cost of concrete compared to conventional thermoelectric materials (e.g., bismuth), thermoelectric applications using concrete may be viable even if the efficiency is not high.

5.4   Materials

The cement matrix is only slightly conductive, with an electrical resistivity of 10^5 or 10^6 Ω.cm [1-3]. By the use of electrically conductive admixtures in the form of particles or short fibers, the resistivity of a cement-based material can be greatly decreased [3]. Continuous fibers can also be used to reduce the resistivity [44], but they cannot be incorporated in a cement mix, and, as a consequence, the making of a continuous fiber cement-based material is much more complicated than that of a short fiber cement-based material. This book only addresses the use of particles and short fibers as electrically conductive admixtures.

Due to the requirements of low material cost and long-term compatibility with the chemical environment in a cement-based material, the electrically conductive admixtures are mainly either steel or carbon. Steel is more conductive than carbon, but it is less available in the form of fine particles or fibers. It is desirable to attain a low resistivity at just a low volume fraction of an admixture, because the workability and compressive strength decrease with increasing volume fraction of the admixture (due to the increase in air void content) [1,2] and the cost increases with the admixture volume fraction. As a consequence, a small particle size, a small fiber diameter and a large aspect ratio are usually attractive for the admixture. However, too small an admixture unit size can be a disadvantage for the conductivity, as shown in Sec. 5.5, due to the electrical contact resistance at the interface between the adjacent admixture units and the large number of such interfaces when the filler unit size is small.

Among the steel admixtures are steel fibers, shavings and dust. Steel shavings and dust are waste materials from the machining of steel components. Although they are less expensive than steel fibers, they are usually less pure and are available in much larger sizes. Steel fibers are available at diameter ranging from 8 μm to 2 mm. Steel fibers that are relatively fine are only available in the form of stainless steel. Stainless steel fibers of diameter 60 μm and 8 μm have received particular attention in relation to electrically conductive cement-based materials, although carbon steel fibers of much larger diameters are typically used in purely structural applications of cement-based materials. The steel fibers used in this section are all made of stainless steel.

Among the carbon admixtures are graphite powder, coke powder (i.e., coke breeze), carbon fibers (typically of diameter ranging from 7 to 15 μm and length around 5 mm) and carbon filaments (typically of diameter ranging from 0.01 to 1 μm and length 100 μm or more). In this chapter, fibers refer to those of diameter 1 μm or above, whereas filaments refer to those of diameter below 1 μm [45]. Carbon fibers are typically made from pitch, polyacrylonitrile or other polymers [46]. Carbon filaments are typically made catalytically from carbonaceous gases [46]. Coke is less crystalline than graphite, so it is less conductive. However, coke is less expensive than graphite. Carbon fibers and filaments are even more expensive than graphite powder. However, their large values of the aspect ratio facilitate electrical connectivity among the conductive admixture units, thereby enhancing the conductivity of the composite. On the other hand, due to the large values of the aspect ratio, fibers and filaments have the tendency to cling together, thus making their dispersion more difficult than the powder counterpart. Dispersion is important for both electrical and mechanical properties of the composites. Very fine particles, such as silica (SiO_2) fume of mean particle size around 0.1 μm, are effective as an admixture for helping fiber dispersion [3,17,18,47].

The various forms of carbon also differ in their mechanical properties and reinforcing effectiveness, which are important for structural materials. Due to its high crystallinity and the consequent tendency to undergo shear between the carbon layers, graphite powder is mechanically weaker than coke powder. On the other hand, carbon fibers and filaments are more effective than the powder for reinforcement, due to their large values of the aspect ratio. The strength is particularly high along the axis of a carbon fiber, due to the preferred orientation of the carbon layers along the fiber axis [46]. For the carbon filaments, the preferred orientation of the carbon layers is not necessarily along the filament axis [46]. A common form of carbon filaments (the form used in cement-matrix composites [48]) has the carbon fibers preferably oriented at an angle to the filament axis. This microstructure is referred to as a fishbone morphology. As a result of the off-axis orientation of the carbon layers in this form of carbon filament, the strength of a filament along its axis is expected to be low, though mechanical testing of a single filament has not been reported.

The reinforcing effectiveness also depends on the bond between admixture and cement. This bond is weak compared to that between filler and polymer (e.g., epoxy) in a polymer-matrix composite. Admixture surface treatment can be used to improve this bond [47,49-51], but the improved bond is still not strong and the surface treatment adds considerably to the cost of the admixture. Thus, a large amount of interface between admixture and cement, as in the case of the admixture being small in unit size (e.g., carbon filaments [48]), can be unattractive for the mechanical properties of the composites.

Intercalation is a chemical process (a reaction) that can increase the conductivity of graphite and crystalline types of carbon fibers or filaments. It involves the insertion of a foreign species (a reactant called an intercalate) between the carbon layers, thereby forming a layered compound called an intercalation compound [52]. Intercalation requires the carbon host to be graphitic. Thus, it cannot occur in the common grades of carbon fiber. Due to the high cost of the crystalline types of carbon fiber, intercalated carbon fibers are expensive. The increase in conductivity is a consequence of the charge transfer between the carbon host and the intercalate. In general, the intercalate can be an electron donor or an electron acceptor. Bromine is an intercalate which is an electron acceptor [52,53]. Intercalation of graphite (a semi-metal) with bromine results in a hole metal. Bromine is a particularly attractive intercalate, due to the stability of the intercalation compound in air after desorption of the part of the intercalate which is loosely held [54].

Graphite powder [55-57], coke powder [56-61], carbon fibers (with intercalation [62] and without intercalation [3,62-65]) and carbon filaments (0.1 μm diameter) [48] have all been used as electrically conductive admixtures in cement-based materials.

5.5   Comparative study

This section provides a comparative review of the effectiveness of carbon and steel admixtures for enhancing the electrical conductivity of cement-based materials. The data used in the comparison were all obtained in the laboratory of the author using the same testing method (i.e., the four-probe method of electrical resistance measurement) and the same specimen configuration [48]. In contrast to the two-probe method, the four-probe method eliminates the resistance of the electrical contacts from the measured resistance and is thus more reliable. Furthermore, all data were obtained on similar materials, i.e., materials that involve Type I portland cement at 28 days of curing without any aggregate. Data on the EMI shielding effectiveness were all obtained in the laboratory of the author using the same testing method (i.e., the coaxial cable method, also called the transfer impedance method, involving the use of a network analyzer and a fixture which serves as an expanded coaxial cable, with the specimen sandwiched at its middle section) and the same annular specimen

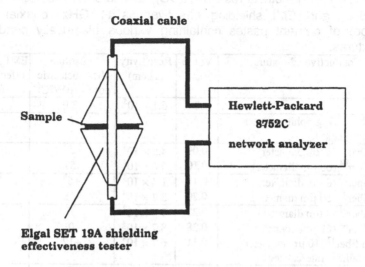

Coaxial cable

Sample

Hewlett-Packard 8752C network analyzer

Elgal SET 19A shielding effectiveness tester

Fig. 5.5 Set-up for measuring the electromagnetic interference shielding effectiveness of various materials.

configuration (Fig. 5.5) [48]. The network analyzer provides the input electromagnetic radiation at a chosen frequency and enables measurement of the attenuation upon transmission (the same as the EMI shielding effectiveness) and the attenuation upon reflection. The attenuation in decibels (dB) is defined as

$$\text{Attenuation (dB)} = 20 \log_{10} (E_i/E), \qquad (5.1)$$

where E_i is the incident field and E is the transmitted or reflected field. Note that $E_i > E$. An electromagnetic wave is a transverse wave, with the electric and magnetic fields perpendicular to the direction of propagation of the wave. Therefore, the electric and magnetic fields in Fig. 5.5 are in the plane of the specimen, while the electromagnetic wave propagates in the direction perpendicular to the plane of the specimen. Please refer to the cited references for details on testing methods and specimen preparation methods.

Although the emphasis of this section is on the attaining of high electrical conductivity in cement-based materials, the EMI shielding effectiveness and the thermoelectric behavior are also addressed, due to their relationship with electrical conductivity. On the other hand, this section does not cover the piezoresistive behavior, which pertains to the effect of strain or stress on the electrical resistivity and is useful for strain/stress sensing [1] (relevant to structural vibration control, traffic monitoring and weighing).

Table 5.1   Electrical resistivity (DC), absolute thermoelectric power (20-65°C) and EMI shielding effectiveness (1 GHz, coaxial cable method) of cement pastes containing various electrically conductive admixtures.

Conductive admixture	Vol.%	Resistivity (Ω.cm)	Absolute thermoelectric power (μV/°C)[a]	EMI shielding effectiveness (dB)
None	0	6.1×10^5	-2.0	4
None, but with graphite powder (<1 μm) coating	/	/	/	14
Steel fiber[43] (8 μm diameter)	0.09	4.5×10^3	/	19
Steel fiber[44] (60 μm diameter)	0.10	5.6×10^4	-57	/
Steel fiber[43] (8 μm diameter)	0.18	1.4×10^3	+5[b]	28
Steel fiber[44] (60 μm diameter)	0.20	3.2×10^4	-68	/
Steel fiber[43] (8 μm diameter)	0.27	9.4×10^2	/	38
Steel fiber[44] (60 μm diameter)	0.28	8.7×10^3	0	/
Carbon fiber[37] (10 μm diameter) (crystalline, intercalated)	0.31	6.7×10^3	+12	/
Steel fiber[43] (8 μm diameter)	0.36	57	/	52
Steel fiber[44] (60 μm diameter)	0.40	1.7×10^3	+20	12[b]
Carbon fiber[37] (10 μm diameter) (crystalline, pristine)	0.36	1.3×10^4	-0.5	/
Steel fiber[43] (8 μm diameter)	0.54	23	/	/
Steel fiber[44] (60 μm diameter)	0.50	1.4×10^3	+26	/
Carbon fiber[37] (15 μm diameter) (amorphous, pristine)	0.48	1.5×10^4	-0.9	/
Carbon filament[24] (0.1 μm diameter)	0.5	1.3×10^4	/	30
Graphite powder[73] (<1 μm)	0.46	2.3×10^5	/	10
Coke powder[33] (< 75 μm)	0.51	6.9×10^4	/	44
Steel fiber[43] (8 μm diameter)	0.72	16	/	59
Steel fiber[43] (8 μm diameter)	0.90	40	/	58
Carbon fiber[37] (15 μm diameter) (amorphous, pristine)	1.0	8.3×10^2	+0.5	15[c]
Carbon fiber[37] (10 μm diameter) (crystalline, intercalated)	1.0	7.1×10^2	+17	/
Carbon filament[24] (0.1 μm diameter)	1.0	1.2×10^4	/	35
Graphite powder[73] (<1 μm)	0.92	1.6×10^5	/	22
Coke powder[33] (< 75 μm)	1.0	3.8×10^4	/	47
Steel dust (0.55 mm)	6.6	/	/	5[b]
Graphite powder[30] (< 45 μm)	37	4.8×10^2	+20	/

[a] Seebeck coefficient (with copper as the reference) minus the absolute thermoelectric power of copper. The Seebeck coefficient (with copper as the reference) is the voltage difference (hot minus cold) divided by the temperature difference (hot minus cold).
[b] Ref. 72.
[c] 0.84 vol.% carbon fiber in cement mortar at 1.5 GHz[74].

Although cement-based materials that are strongly piezoresistive (Sec. 2.6) are also electrically conductive, the relationship between piezoresistivity and conductivity is weak [1,66,67]. Piezoresistivity, which pertains to the conductive behavior, is distinct from piezoelectricity, which pertains to the dielectric behavior [68].

Steel rebars used to reinforce concrete are electrically conductive. Consequently they increase the conductivity of concrete. However, they do not affect the EMI shielding effectiveness of concrete, due to their large dimensions and the high frequency of the electromagnetic radiation. The effect of steel rebars is beyond the scope of this section. However, the effect of steel fibers is addressed in this section.

Table 5.1 compares the effectiveness of various electrically conductive admixtures at similar volume fractions in cement paste (without aggregate, whether fine or coarse). Among the various carbon and steel admixtures, stainless steel fibers (8 μm diameter) [69] are most effective for decreasing the DC electrical resistivity and for providing EMI shielding. The lowest resistivity attained in Table 5.1 is 16 Ω.cm; the highest shielding effectiveness attained in Table 5.1 is 59 dB.

The resistivity decreases monotonically, and the EMI shielding effectiveness increases monotonically as the conductive admixture content increases for any given admixture, except that steel fiber (8 μm diameter) gives lower resistivity at 0.72 than 0.90 vol.%. This exception is presumably a consequence of the increase in air void content with increasing fiber volume fraction.

Stainless steel fiber of diameter 60 μm [38,70,71] at essentially the same volume fraction as steel fiber of diameter 8 μm gives much higher resistivity. However, steel fiber of diameter 60 μm gives the highest magnitude of the absolute thermoelectric power (Sec. 5.6.1), which reaches –68 μV/°C. The absolute thermoelectric power of steel fiber (60 μm diameter) cement does not vary monotonically with increasing fiber volume fraction, although the resistivity decreases monotonically with increasing fiber volume fraction [70]. This suggests that carrier scattering at interfaces (e.g., interface between steel and cement) probably dominates the origin of the thermoelectric behavior of steel-cement composites [71]. This suggestion is supported by the opposite signs of the absolute thermoelectric power of the steel fiber itself (+4 μV/°C [70]) and the cement matrix (-2 μV/°C [41]).

Among the three steel admixtures, the effectiveness for shielding decreases in the order: steel fiber (8 μm diameter) [43], steel fiber (60 μm diameter) [44] and steel dust (0.55 mm) [72], as shown by comparing the shielding performance of steel fiber (8 μm diameter) at 0.36 vol.%, steel fiber (60 μm diameter) at 0.40 vol.%, and steel dust (0.55 mm) at 6.6 vol.%. Thus, the greater the unit size of the steel admixture, the less effective the admixture for shielding. This is expected from the skin effect, which refers to the phenomenon in which electromagnetic radiation at a high frequency penetrates only the near surface region of a conductor. The skin depth refers to the depth in the material at which the amplitude of the electromagnetic wave is diminished to 1/e of the incident value. The higher the frequency of the wave, the smaller the skin depth. The higher the electrical conductivity of the materal, the smaller the skin depth.

Among the carbon admixtures, carbon fiber (15 μm diameter) is most effective for decreasing the resistivity. Intercalation does not decrease the resistivity of the composite much, but it greatly enhances the thermoelectric behavior [42]. This is due to the role of carrier hopping across the fiber-cement interface in governing the electrical conduction below the percolation threshold and the increase in the activation energy of the hopping (as determined from the variation of the resistivity with temperature) upon intercalation [62].

Carbon filament (0.1 μm diameter) is much less effective than carbon fiber (15 μm diameter) for lowering the resistivity, but is much more effective for providing EMI shielding. This is because of (i) the skin effect, which makes an admixture with a smaller unit size more effective for shielding at the same volume fraction, and (ii) the contact resistance at the admixture-cement interface, which is large in area per unit volume when the admixture unit size is small.

Coke powder is less effective than carbon filament for lowering the resistivity, due to its particulate (nonfibrous) nature, but is more effective than carbon filament for providing shielding (presumably due to better dispersion). The low cost of coke adds to the attraction of coke-cement for EMI shielding.

Graphite powder (< 1 μm in particle size, the solid portion of a water-based colloid) [73] is less effective than carbon fiber (whether amorphous or crystalline), carbon filament or coke powder for lowering the resistivity. Its inferiority to carbon fiber and filament is due to its particulate nature; its inferiority to coke powder is probably related to its small particle size and the fact that its volume fractions are below the percolation threshold. The graphite powder is inferior to carbon filament and coke powder for shielding, but it is superior to carbon fiber for shielding. The inferiority of graphite powder to carbon filament for shielding is due to its particulate nature. The origin of the inferiority to coke powder is presently not clear, as the larger particle size of coke is expected to be disadvantageous for shielding. The superiority of graphite powder to carbon fiber for shielding is due to its small particle size and the skin effect.

Graphite powder (< 45 μm) is less effective than carbon fiber (whether amorphous or crystalline) for lowering the resistivity, as 37 vol.% of the graphite powder and 1.0 vol.% of carbon fiber have similar effects. This is due to the particulate nature of the graphite.

All the carbon and steel admixtures at all volume fractions investigated in terms of the thermoelectric behavior cause the absolute thermoelectric power to be less negative (more positive), except that steel fiber (60 μm diameter) up to 0.2 vol.% causes the absolute thermoelectric power to be more negative (as negative as –68 μV/°C). The steel fiber (60 μm diameter) itself, without cement, has a positive value (+4 μV/°C) of the absolute thermoelectric power [70], but it can cause the absolute thermoelectric power of a cement paste to be more negative or more positive, depending on its volume fraction. The fact that bromine intercalation of carbon fiber causes the absolute thermoelectric power to be much more positive [43] supports the notion that holes contribute to the thermoelectric behavior of the carbon-cement composites. However, both electrons (from steel) and carrier scattering (at the steel-cement interface) probably contribute to the thermoelectric behavior of the steel-cement composite.

For providing a large magnitude of the absolute thermoelectric power, the effectiveness of steel fiber (60 μm diameter) is outstanding, if it is used up to 0.2 vol.% only. Beyond 0.2 vol.%, the absolute thermoelectric power becomes more positive as the fiber volume fraction increases. For thermoelectric energy generation, materials with opposite signs of the absolute thermoelectric power are usually connected in series in order to have an additive effect on the voltage generated. Thus, materials that exhibit strongly positive and strongly negative values of the absolute thermoelectric power are both useful. Steel fiber (60 μm diameter) at 0.20 vol.% gives the most negative value (-68 μV/°C), whereas the same fiber at 0.50 vol.% gives the most positive value (+26 μV/°C). Graphite powder at 37 vol.% and steel fiber (60 μm diameter) at 0.40 vol.% give the same value of +20 μV/°C, but the high volume fraction of graphite powder is unattractive for the mechanical properties.

The following rank ordering of the effectiveness of various conductive admixtures were obtained from Table 5.1 by comparing the effectiveness at similar volume fractions of the admixtures. For lowering the resistivity, the effectiveness of the various admixtures decreases in the order: steel fiber (8 μm diameter), steel fiber (60 μm diameter), carbon fiber (15 μm diameter), carbon filament (0.1 μm diameter), coke powder (< 75 μm) and graphite powder (< 1 μm). Steel fiber (8 μm diameter) is exceptionally effective compared to the other admixtures. Consistent with the exceptionally low resistivity of steel fiber (8 μm diameter) cement is the exceptionally high effectiveness of the material for use as a resistance heating element [22]. In particular, the effectiveness is higher than that of carbon fiber (15 μm diameter) cement [22]. The steel fiber cement has not been evaluated for use as an overlay for cathodic protection, but carbon fiber mortar has been shown to be more effective than plain mortar (without fiber) for cathodic protection, as the required voltage is reduced [30].

For providing EMI shielding, the effectiveness of the various admixtures decreases in the order: steel fiber (8 μm diameter), coke powder (< 75 μm), carbon filament (0.1 μm diameter), graphite powder (< 1 μm), steel fiber (60 μm diameter), carbon fiber (15 μm diameter) and steel dust (0.55 mm). That steel fiber (60 μm diameter) is better than carbon fiber (15 μm diameter) is just suggested by the slightly lower shielding effectiveness attained by steel fiber (60 μm diameter, 0.40 vol.%) than carbon fiber (15 μm diameter, 1.0 vol.%). The relative shielding performance of graphite powder (< 1 μm) and steel fiber (60 μm diameter) has not been investigated, though both are included in the ranking above. The other parts of the rank ordering are well substantiated. Although steel fiber (8 μm diameter) is more effective than coke powder for shielding, it is much more expensive than coke powder. Thus, coke powder is preferred when both cost and performance are considered.

A material which is superior in conductivity is not necessarily also superior in shielding, as shown by comparing carbon filament cement and coke cement. Carbon filament (0.5 vol.%) cement is better than coke (0.5 vol.%) cement in conductivity, but is inferior in shielding. A material which is stronger in its thermoelectric behavior is not necessarily better in conductivity, as shown by comparing steel fiber (8 μm diameter) cement and steel fiber (60 μm diameter) cement. Steel fiber (60 μm diameter, 0.2 vol.%) cement is stronger thermoelectrically, but is less conductive than steel fiber (8 μm diameter, 0.2 vol.%) cement. A material which is superior in shielding is not necessarily also superior in conductivity, as shown by comparing graphite powder (< 1 μm) cement and carbon fiber (15 μm diameter) cement. The former (0.92 vol.%) is better than the latter (1.0 vol.%) for shielding, but is much inferior in conductivity.

The use of graphite powder (< 1 μm) in the form of a water-based colloid to coat cement without admixture results in an increase of the shielding effectiveness from 4 to 14 dB [73], as shown in Table 5.1. The coating thickness is 0.3 mm. When the substrate is Mylar (electromagnetically transparent) instead of cement, the shielding effectiveness is 11 dB [73]. Although the coating method is effective to a limited degree for shielding and is convenient for implementation in existing structures, it suffers from the tendency to be damaged by abrasion and wear.

Fly ash as an admixture results in a negligible decrease of the electrical resistivity, but it enhances the shielding effectiveness of cement paste from 4 to 8 dB at 1 GHz when the fly ash : cement weight ratio is increased from 0 : 100 to 100 : 0 [75]. This is attributed to the Fe_2O_3 component (15.4 wt.%) in the fly ash. In contrast, silica fume, which has only <0.5 wt.% Fe_2O_3, enhances the shielding effectiveness of cement paste negligibly, though it decreases the resistivity more significantly than fly ash [75].

Table 5.2 Seebeck coefficient.*

Material	Temperature (°C)	Seebeck coefficient (μV/K)
Al	100	-0.20
Cu	100	+3.98
W	100	+3.68
ZnSb	200	+220
Ge	700	-210
$Bi_2Te(Se)_3$	100	-210
TiO_2	725	-200

* M. Ohring, *Engineering Materials Science*, Academic Press, San Diego, 1995, pp. 633.

5.6 Thermoelectric behavior

5.6.1 Introduction

Thermoelectric phenomena involve the transfer of energy between electric power and thermal gradients. They are widely used for cooling and heating, including air conditioning, refrigeration, thermal management and the generation of electrical power from waste heat.

The thermoelectric phenomenon involving the conversion of thermal energy to electrical energy is embodied in the Seebeck effect, i.e., the greater concentration of carrier above the Fermi energy at the hot point than the cold point, the consequent movement of mobile carrier from the hot point to the cold point and the resulting voltage difference (called the Seebeck voltage) between the hot and cold points. If the mobile carrier is electrons, the hot point is positive in voltage relative to the cold point. If the mobile carrier is holes, the cold point is positive relative to the hot point. Hence, a temperature gradient results in a voltage. The change in Seebeck voltage (hot minus cold) per degree C temperature rise (hot minus cold) is called the thermoelectric power, the thermopower, or the Seebeck coefficient. Table 5.2 gives the values of the Seebeck coefficient of various materials. The Seebeck effect is the basis for thermocouples.

For the thermoelectric phenomenon (Peltier effect) which involves the conversion of electrical energy to thermal energy (for heating or cooling), thecombination of a low thermal conductivity (to reduce heat transfer loss), a high electrical conductivity (to reduce Joule heating) and a high thermoelectric power is required. These three factors are combined in the thermoelectric figure of merit Z, which is defined as

Table 5.3   Thermoelectric figure of merit Z*

Materials that form junction	$Z\ (10^{-3}\ K^{-1})$
Chromel-constantan	0.1
Sb-Bi	0.18
ZnSb-constantan	0.5
PbTe(p)-PbTe(n)	1.3
$Bi_2Te_3(p)$-$Bi_2Te_3(n)$	2.0

* M. Ohring, *Engineering Materials Science*, Academic Press, San Diego, 1995, pp. 633.

$$Z = \frac{\alpha_{AB}^2}{\left[\left(\frac{\kappa}{\sigma}\right)_A^{1/2} + \left(\frac{\kappa}{\sigma}\right)_B^{1/2}\right]^2}, \qquad (5.2)$$

where A and B are the two dissimilar conductors that form a junction, α_{AB} is the Seebeck coefficient difference ($\alpha_{AB} = \alpha_A - \alpha_B$), κ is the thermal conductivity and σ is the electrical conductivity. Values of Z for various junctions are shown in Table 5.3. A junction commonly involves a p-type semiconductor and an n-type semiconductor, as in the last two entries in Table 5.3. In practice a current is passed through the junction in order to attain either heating or cooling. A change in current direction causes a change from heating to cooling, or vice versa.

Thermoelectric behavior has been observed in metals, ceramics and semiconductors, as they are electrically conducting. Composite engineering provides a route to develop better thermoelectric materials, as composites with different properties can be combined in a composite in order to achieve a high figure of merit. Metals are usually high in both thermal and electrical conductivities. Since the combination of low thermal conductivity and high electrical conductivity is not common in single-phase materials, the composite route is valuable. Moreover, the composite route can be used to enhance the mechanical properties.

5.6.2   Cement-based materials for the Seebeck effect

The Seebeck effect provides a renewable source of energy, in addition to providing the basis for thermocouples, which are used for temperature measurement. Moreover, it is relevant to the reduction of environmental pollution and global warming.

The approach used in this section involves taking structural composites as a starting point and modifying these composites for the purpose of enhancing

Table 5.4   Properties of steel fibers

Fiber diameter	60 μm
Tensile strength	970 MPa
Tensile modulus	200 GPa
Elongation at break	3.2%
Electrical resistivity	6×10^{-5} Ω.cm
Specific gravity	7.7 g cm^{-3}

the thermoelectric properties. In this way, the resulting composites are multifunctional (i.e., both structural and thermoelectric). Because structural composites are used in large volumes, the rendering of the thermoelectric function to these materials means the availability of large volumes of thermoelectric materials for use in, say, electrical energy generation. Moreover, temperature sensing is useful for structures for the purpose of thermal control, energy saving and hazard mitigation. The rendering of the thermoelectric function to a structural material also means that the structure can sense its own temperature without the need for embedded or attached thermometric devices, which suffer from high cost and poor durability. Embedded devices, in particular, cause degradation of the mechanical properties of the structure.

A route used in the tailoring involves the choice of fibers. It impacts the thermoelectric properties in any direction for a composite containing randomly oriented fibers.

Cement is a low-cost, mechanically rugged and electrically conducting material which can be rendered n-type or p-type by the use of appropriate admixtures, such as short carbon fibers (which contribute holes) for attaining p-type cement and short steel fibers for attaining n-type cement [39-42]. (Cement itself is weakly n-type in relation to electronic/ionic conduction [39]). The fibers also improve the structural properties, such as increasing the flexural strength and toughness and decreasing the drying shrinkage [76-83]. Furthermore, cement-based junctions can be easily made by pouring the dissimilar cement mixes side by side.

The steel fibers used to provide strongly n-type cement paste were made of stainless steel No. 434, as obtained from International Steel Wool Corp. (Springfield, OH). The fibers were cut into pieces of length 5 mm prior to use in the cement paste in the amount of 0.5% by mass of cement (i.e., 0.10 vol.%). The properties of the steel fibers are shown in Table 5.4. The mechanical properties of mortars containing these fibers are described in Ref. 77. However, no aggregate, whether coarse or fine, was used in the thermoelectric investigation.

The carbon fibers used to provide p-type cement paste were isotropic pitch based, unsized, and of length ~ 5 mm, as obtained from Ashland Petroleum

Table 5.5   Properties of carbon fibers

Filament diameter	$15 \pm 3 \ \mu m$
Tensile strength	690 MPa
Tensile modulus	48 GPa
Elongation at break	1.4%
Electrical resistivity	$3.0 \times 10^{-3} \ \Omega.cm$
Specific gravity	$1.6 \ g \ cm^{-3}$
Carbon content	98 wt.%

Co. (Ashland, KY). They were used in the amount of either 0.5% or 0.1% by mass of cement (i.e., either 0.48 or 0.96 vol.% in the case of cement paste with silica fume, and either 0.41 or 0.82 vol.% in the case of cement paste with latex). Silica fume, due to its fine particulate nature, is particularly effective for enhancing the fiber dispersion [3,84]. The fiber properties are shown in Table 5.5. No aggregate (fine or course) was used. The cement paste with carbon fibers in the amount of 1.0% by mass of cement was p-type, whereas that with carbon fibers in the amount of 0.5% by mass of cement was slightly n-type, as shown by thermoelectric power measurement [39].

The cement used in all cases was portland cement (Type I) from Lafarge Corp. (Southfield, MI). Silica fume (Elkem Materials, Inc., Pittsburgh, PA, EMS 965) was used in the amount of 15% by mass of cement. The methylcellulose, used in the amount of 0.4% by mass of cement, was Dow Chemical Corp., Midland, MI, Methocel A15-LV. The defoamer (Colloids, Inc., Marietta, GA, 1010) used whenever methylcellulose was used was in the amount of 0.13 vol.%. The latex, used in the amount of 20% by mass of cement, was a styrene butadiene copolymer (Dow Chemical Co., Midland, MI, 460NA) with the polymer making up about 48% for the dispersion and with the styrene and butadiene having a mass ratio of 66:34. The latex was used along with an antifoaming agent (Dow Corning Corp., Midland, MI, No. 2410, 0.5% by mass of latex).

A rotary mixer with a flat beater was used for mixing. Methylcellulose (if applicable) was dissolved in water, and then the defoamer was added and stirred by hand for about 2 min. Latex (if applicable) was mixed with the antifoam by hand for about 1 min. Then the methylcellulose mixture (if applicable), the latex mixture (if applicable), cement, water, silica fume (if applicable), carbon fibers (if applicable) and steel fibers (if applicable) were mixed in the mixer for 5 min.

A junction between any two types of cement mix was made by pouring the two different mixes into a rectangular mold (160 x 40 x 40 mm) separately, such that the time between the two pours was 10-15 min. The two mixes were poured into two side-by-side compartments of the mold, and the paper (2 mm thick, without oil on it) separating the compartments was removed immediately after the completion of the two pours. Each compartment was roughly half the length of the entire mold.

After pouring into oiled molds, an external electrical vibrator was used to facilitate compaction and decrease the amount of air bubbles. The resulting junction could be seen visually, due to the color difference between the two halves of a sample. The samples were demolded after 1 day and cured in air at room temperature (relative humidity = 100%) for 28 days.

Five types of cement paste were prepared, namely (i) plain cement paste (weakly n-type, consisting of just cement and water), (ii) steel fiber cement paste (strongly n-type, consisting of cement, water and 60 μm diameter steel fibers), (iii) carbon fiber silica fume cement paste (very weakly n-type, consisting of cement, water silica fume, methylcellulose, defoamer and carbon fibers in the amount of 0.5% by mass of cement), (iv) carbon fiber silica fume cement paste (p-type, consisting of cement, water, silica fume, methylcellulose, defoamer and carbon fibers in the amount of 1.0% by mass of cement), and (v) carbon fiber latex cement paste (very weakly n-type, consisting of cement, water, latex and carbon fibers). The water/cement ratio was 0.45 for pastes (i), (ii), (iii) and (iv), and was 0.25 for paste (v). The absolute thermoelectric power of each paste is shown in Table 5.6 [39,42].

Three pairs of cement paste were used to make junctions, as described in Table 5.7. Thermocouple testing was conducted by heating the junction by resistance heating, which was provided by nichrome heating wire (wound around the whole perimeter of the sample over a width of 10 mm that was centered at the junction), a transformer and a temperature controller. The voltage difference

Table 5.6   Absolute thermoelectric power (μV/°C)

Cement paste	Volume fraction fibers	μV/°C	Type	Ref.
(i) Plain	0	1.99 ± 0.03	weakly n	41
(ii) $S_f (0.5^*)$	0.10%	53.3 ± 4.8	strongly n	42
(iii) $C_f (0.5^*) + SF$	0.48%	0.89 ± 0.09	weakly n	41
(iv) $C_f (1.0^*) + SF$	0.95%	-0.48 ± 0.11	p	41
(v) $C_f (0.5^*) + L$	0.41%	1.14 ± 0.05	weakly n	41

Note: SF = silica fume; L = latex.

Table 5.7   Cement junctions

Junction	Pastes involved	Junction type	Thermocouple sensitivity (μV/°C)	
			Heating	Cooling
(a)	(iv) and (ii)	pn	70 ± 7	70 ± 7
(b)	(iii) and (ii)	nn$^+$	65 ± 5	65 ± 6
(c)	(v) and (ii)	nn$^+$	59 ± 7	58 ± 5

Note: nn$^+$ refers to a junction between a weakly n-type material and a strongly n-type material.

between the two ends of a sample was measured by using electrical contacts in the form of copper wire wound around the whole perimeter of the sample at each end of the sample. Silver paint was present between the copper wire and the sample surface under the wire. The copper wires from the two ends were fed to a Keithley 2001 multimeter for voltage measurement. A T-type thermocouple was positioned to almost touch the heating wire at the junction. Another T-type thermocouple was attached to one of the two ends of the sample (at essentially room temperature). The difference in temperature between these two locations governs the voltage. Voltage and temperature measurements were done simultaneously using the multimeter, while the junction temperature was varied through resistance heating. The voltage difference divided by the temperature difference yielded the thermocouple sensitivity.

Fig. 5.6 [38] shows plots of the thermocouple voltage versus the temperature difference (relative to essentially room temperature) for junction (a). The thermocouple voltage increases monotonically and reversibly with increasing temperature difference for all junctions. The thermocouple voltage noise decreases and the thermocouple sensitivity (Table 5.7 [38]) and reversibility increase in the order: (c), (b) and (a). The highest thermocouple sensitivity is 70 ± 7 μV/°C, as attained by junction (a) both during heating and cooling. This value approaches that of commercial thermocouples. That junction (a) gives the best thermocouple behavior (in terms of sensitivity, linearity, reversibility and signal-to-noise ratio) is due to the greatest degree of dissimilarity between the materials that make up the junction. The linearity of the plot of thermocouple voltage versus temperature difference is better during cooling than during heating for junction (a). Being a pn junction, junction (a) exhibits a diode-like current-voltage characteristic (Fig. 5.7), although the current rectification is not perfect [37,38].

The values of the thermocouples sensitivity (Table 5.7) are higher than (theoretically equal to) the difference in the absolute thermoelectric power of the corresponding two cement pastes that make up the junction (Table 5.6). For example, for junction (a), the difference in the absolute thermoelectric power of

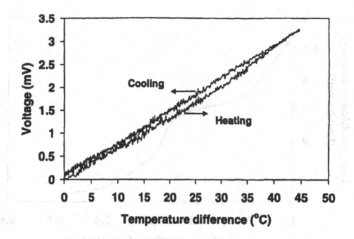

Fig. 5.6   Variation of the cement-based thermocouple voltage with temperature difference during heating and then cooling for junction (a) of Table 5.7.

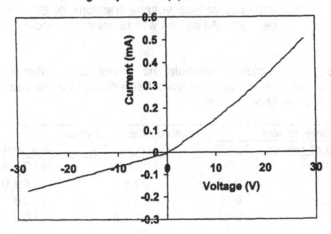

Fig. 5.7   Current-voltage characteristic of junction (a) of Table 5.7.

pastes (iv) and (ii) is 54 μV/°C, but the thermocouple sensitivity is 70 μV/°C. The reason for this is unclear, but may be related to an electrochemical effect. Nevertheless, a higher thermocouple sensitivity does correlate with a greater difference in the absolute thermoelectric power.

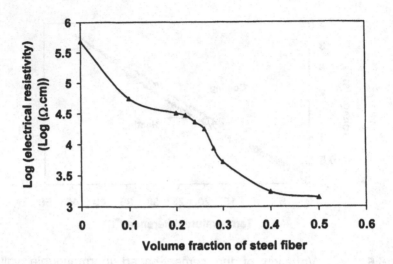

Fig. 5.8   Volume electrical resistivity (log scale) of cement pastes containing various volume fractions of 60 μm diameter steel fiber. All pastes with fibers contained silica fume.

Table 5.8   Absolute thermoelectric power and volume electrical resistivity of cement pastes (with silica fume except for the paste without fiber) and of steel fiber by itself.

Fiber content		Absolute thermoelectric	
% by mass of cement	Vol. %	power (μV/°C)[†]	Resistivity (Ω.cm)
0	0	-1.99 ± 0.04	$(4.7 \pm 0.4) \times 10^5$
0.5	0.10	-57 ± 4	$(5.6 \pm 0.5) \times 10^4$
1.0	0.20	-68 ± 5	$(3.2 \pm 0.3) \times 10^4$
1.1	0.22	-48 ± 5	$(3.0 \pm 0.2) \times 10^4$
1.2	0.24	-25 ± 2	$(2.3 \pm 0.2) \times 10^4$
1.3	0.26	-13 ± 1	$(1.8 \pm 0.1) \times 10^4$
1.4	0.28	0.0 ± 0.2	$(8.7 \pm 0.1) \times 10^3$
1.5	0.30	+6.3 ± 1.2	$(5.3 \pm 0.4) \times 10^3$
2.0	0.40	+20 ± 3	$(1.7 \pm 0.1) \times 10^3$
2.5	0.50	+26 ± 3	$(1.4 \pm 0.2) \times 10^3$
/	100*	+3.76 ± 0.15	6×10^{-5} §

* Steel fiber by itself.
† Measured during heating.
§ From the manufacturer's data sheet.

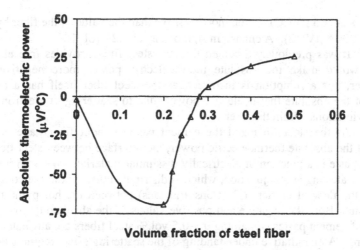

Fig. 5.9   Absolute thermoelectric power of cement pastes
                   containing various volume fractions of 60 μm diameter
                   steel fiber. All pastes with fibers contained silica fume.

        Table 5.8 [70] and Fig. 5.8 [70] show that the volume electrical
resistivity of cement paste is decreased monotonically by steel fiber addition.
The higher fiber volume fraction, the lower is the resistivity. The absence of an
abrupt drop in resistivity as the fiber content increases suggests that all of the
fiber volume fractions used are below the percolation threshold, as expected
from the previously reported percolation threshold of 0.5-1.0 vol.% for carbon
fiber (15 μm diameter) cement paste [3].
        Table 5.8 [70] and Fig. 5.9 [70] give the absolute thermoelectric power
of cement pastes and of the steel fiber by itself. The steel fiber itself has a
positive value of the absolut thermoelectric power, whereas cement paste
without fiber has a negative value [71]. The addition of fibers up to 0.20 vol.%
makes the value more negative, as reported for the case without silica fume [42].
At the same fiber volume fraction of 0.10%, the use of silica fume changes the
absolute thermoelectric power from –53 ± 5 μV/°C [42] to –57 ± 4 μV/°C [70],
due to a higher degree of fiber dispersion [3], as shown by the decrease in
electrical resistivity from 8×10^4 to 6×10^4 Ω.cm [42]. Increase of the fiber
volume fraction from 0.10% to 0.20% causes the absolute thermoelectric power
to become even more negative, reaching –68 ± 5 μV/°C, which is the highest in
magnitude among all cement pastes studied. However, increase of the fiber
content beyond 0.20 vol.% makes the value less negative and more positive (as

high as + 26 ± 3 µV/°C) – even more positive than the value of the fiber by itself (+3.76 ± 0.15 µV/°C). A change in sign occurs at 0.28 vol.%.

It was previously assumed that the steel fiber provides free electrons which would make the absolute thermoelectric power more negative [42]. However, this assumption is incorrect, as the steel fiber itself has a positive value of the absolute thermoelectric power (due to scattering of electrons from lattice vibrations within the fiber).

As the steel fiber and the cement paste without fiber have opposite signs of the absolute thermoelectric power, the interface between steel fiber and cement paste is a junction of electrically dissimilar materials, like a pn junction. Carrier scattering at this junction, which is distributed throughout the composite, affects the flow of carriers (electrons and ions) between the hot point and the cold point. Both the negative and positive values of the absolute thermoelectric power of cement pastes containing 0.1-0.5 vol.% steel fibers are attributed to the scattering. A quantitative understanding of the scattering effect requires detailed information on the mean free path and mean free time of the carriers.

The situation is quite different in the case of carbon fiber cement paste. The carbon fiber contributes to hole conduction [39-41], thus making the absolute thermoelectric power of the cement-matrix composite more positive [39]. By using intercalated carbon fiber which provides even more holes, the absolute thermoelectric power becomes even more positive [43]. Thus, hole conduction rather than scattering dominates the origin of the Seebeck effect in carbon fiber cement paste. In contrast, scattering rather than electron conduction dominates the Seebeck effect in steel fiber cement paste.

5.7   Conclusion

Electrically conductive cement-based materials are useful for electrical grounding, lightning protection, resistance heating, static charge dissipation, electromagnetic interference (EMI) shielding, cathodic protection, and thermoelectric energy generation. The science and applications of electrically conductive cement-based materials are reviewed in this chapter. In addition, a comparative study of the effectiveness of various electrically conductive admixtures (discontinuous forms of steel and carbon) for lowering the electrical resistivity of cement shows that the effectiveness decreases in the order: steel fiber of the diameter 8 µm, steel fiber of diameter 60 µm, carbon fiber of diameter 15 µm, carbon filament of diameter 0.1 µm, coke powder (< 75 µm) and graphite powder (< 1 µm). For EMI shielding, the effectiveness decreases in the order: steel fiber of diameter 8 µm, coke powder (< 75 µm), carbon filament of diameter 0.1 µm, graphite powder (< 1 µm), steel fiber of diameter 60 µm, carbon fiber of diameter 15 µm, and steel dust of size 0.55 mm. By using steel fiber (8 µm diameter) at 0.72 vol.%, a resistivity of 16 Ω.cm and an EMI shielding effectiveness of 59 dB (1 GHz) were attained. The carbon admixtures cause the absolute thermoelectric power to be more positive, whereas the steel admixtures can cause the absolute thermoelectric power to be more positive or more negative. In particular, steel fiber of diameter 60 μm at 0.2 vol.% causes the absolute thermoelectric power to be strongly negative (-68 μV/°C).

References

1. D.D.L. Chung, *J. Mater. Sci.* 36, 1315-1324 (2001).
2. D.D.L. Chung, *Appl. Thermal Eng.* 21(ER16), 1607-1619 (2001).
3. P.-W. Chen and D.D.L. Chung, *J. Electron. Mater.* 24(1), 47-51 (1995).
4. D.D.L. Chung, *J. Mater. Eng. Perf.* 11(2), 194-204 (2002).
5. X. Wang, Y. Wang and Z. Jin, *J. Mater. Sci.* 37(1), 223-227 (2002).
6. X. Fu and D.D.L. Chung, *Cem. Concr. Res.* 25(4), 689-694 (1995).
7. M.S. Morsy, *Cem. Concr. Res.* 29, 603-606 (1999).
8. J.G. Wilson and N.K. Gupta, *Build. Res. Info.* 24(4), 209-212 (1996).
9. S.A. Abo El-Enein, M.F. Kotkata, G.B. Hanna, M Saad and M.M. Abd El Razek, *Cem. Concr. Res.* 25(8), 1615-1620 (1995).
10. H.C. Kim, S.Y. Kim and S.S. Yoon, *J. Mater. Sci.* 30(15), 3768-3772 (1995).
11. S. Wen and D.D.L. Chung, *Cem. Concr. Res.* 29(6), 961-965 (1999).
12. X. Fu and D.D.L. Chung, *Cem. Concr. Res.* 26(7), 985-991 (1996).
13. P.J. Tumidajski, *Cem. Concr. Res.* 26(4), 529-534 (1996).
14. G. Ping, X. Ping and J.J. Beaudoin, *Cem. Conc. Res.* 23(3), 581-591 (1993).
15. T.O. Mason, S.J. Ford, J.D. Shane, J.-H. Hwang and D.D. Edwards, *Adv. Cem. Res.* 10(4), 143-150 (1998).
16. D.E. MacPhee, D.C. Sinclair and S.L. Stubbs, *J. Mater. Sci. Lett.* 15(18), 1566-1568 (1996).
17. J. Cao and D.D.L. Chung, *Cem. Concr. Res.* 31(11), 1633-1637 (2001).
18. P.-W. Chen, X. Fu and D.D.L. Chung, *ACI Mater. J.* 94(2), 147-155 (1997).
19. J. Cao and D.D.L. Chung, *Cem. Concr. Res.* 31(2), 309-311 (2001).
20. X. Fu and D.D.L. Chung, *Composite Interfaces* 4(4), 197-211 (1997).
21. D.D.L. Chung, *Composite Interfaces* 8(1), 67-82 (2001).
22. S. Wang, S. Wen and D.D.L. Chung, *Cem. Concr. Res.*, in press.
23. S.A. Yehia and C.Y. Tuan, *Concr. Int.* 24(2), 56-60 (2002).
24. M. Unz, *Corrosion* 16, 123 (1960).
25. B. Heuze, *Mater. Protection* 4(11), 57 (1965).
26. D.A. Hausmann, *Mater. Protection* 8(10), 66 (1969).
27. G. Baronio, M. Berra, L. Bertolini and T. Pastore, *Cem. Concr. Res.* 26(5), 683 (1996).

28.   D.D.L. Chung, *J. Mater. Eng. Perf.* 9(5), 585-588 (2000).
29.   R. Pangrazzi, W.H. Hartt and R. Kessler, *Corrosion* (Houston) 50(3), 186 (1994).
30.   J. Hou and D.D.L. Chung, *Cem. Concr. Res.* 27(5), 649-656 (1997).
31.   B.D. Mottahed and S. Manoocheheri, *Polym.-Plastics Tech. Eng.* 34(2), 271-346 (1995).
32.   P.S. Neelakanta and K. Subramaniam, *Adv. Mater. Processes* 141(3), 20-25 (1992).
33.   G. Lu, X. Li and H. Jiang, *Compos. Sci. Tech.* 56, 193-200 (1996).
34.   A. Kaynak, A. Polat and U. Yilmazer, *Mater. Res. Bull.* 31(10), 1195-1203 (1996).
35.   D.D.L. Chung, *J. Mater. Eng. Perf.* 9(3), 350-354 (2000).
36.   X. Fu and D.D.L. Chung, *Cem. Concr. Res.* 28(6), 795-801 (1998).
37.   S. Wen and D.D.L. Chung, *Cem. Concr. Res.* 31(3), 507-510 (2001).
38.   S. Wen and D.D.L. Chung, *J. Mater. Res.* 16(7), 1989-1993 (2001).
39.   M. Sun, Z. Li, Q. Mao and D. Shen, *Cem. Concr. Res.* 28(4), 549-554 (1998).
40.   M. Sun, Z. Li, Q. Mao and D. Shen, *Cem. Concr. Res.* 28(12), 1707-1712 (1998).
41.   S. Wen and D.D.L. Chung, *Cem. Concr. Res.* 29(12), 1989-1993 (1999).
42.   S. Wen and D.D.L. Chung, *Cem. Concr. Res.* 30(4), 661-664 (2000).
43.   S. Wen and D.D.L. Chung, *Cem. Concr. Res.* 30(8), 1295-1298 (2000).
44.   S. Wen, S. Wang and D.D.L. Chung, *J. Mater. Sci.* 35(14), 3669-3675 (2000).
45.   D.D.L. Chung, *Carbon* 39(8), 1119-1125 (2001).
46.   D.D.L. Chung, Carbon Fiber Composites, Butterworth-Heinemann, 1994.
47.   D.D.L. Chung, *J. Mater. Sci.* 37(4), 673-682 (2002).
48.   X. Fu and D.D.L. Chung, *Carbon* 36(4), 459-462 (1998).
49.   X. Fu, W. Lu and D.D.L. Chung, *Carbon* 36(9), 1337-1345 (1998).
50.   Y. Xu and D.D.L. Chung, *ACI Mater. J.* 97(3), 333-342 (2000).
51.   X. Fu and D.D.L. Chung, *ACI Mater. J.* 94(3), 203-208 (1997).
52.   D.D.L. Chung, Graphite intercalation compounds. *Encyclopedia of Materials: Science and Technology,* K.H.J. Buschow, R.W. Cahn, M.C. Flemings, B. Ilschner, E.J. Kramer and S. Mahajan (eds.), Elsevier, Oxford, 4, 3641-3645 (2001).
53.   D.D.L. Chung, *Phase Transitions* 8, 35-57 (1986).
54.   C.T. Ho and D.D.L. Chung, *Carbon* 28(6), 825-830 (1990).
55.   S. Wen and D.D.L. Chung, *Carbon* 40(13), 2495-2497 (2002).
56.   P.L. Zaleski, D.J. Derwin and W.H. Flood Jr., U.S. Patent 5,707,171 (1998).
57.   P. Xie, P. Gu, Y. Fu and J.J. Beaudoin, U.S. Patent 5,447,564 (1995).

58. J. Cao and D.D.L. Chung, *Carbon*, in press.
59. G.H. Anderson, *Proceedings of the Conference on cathodic protection of reinforced concrete bridge decks*, 82-88 (1985).
60. K.C. Clear, *Proceedings of the Conference on cathodic protection of reinforced concrete bridge decks*, 55-65 (1985).
61. P. Xie and J.J. Beaudoin, *ACI SP 154-21*, Advances in concrete technology, Ed. V.M. Malhotra, 399-417 (1995).
62. S. Wen and D.D.L. Chung, *Carbon* 39, 369-373 (2001).
63. D.D.L. Chung, *Compos.: Part B* 31(6-7), 511-526 (2000).
64. J. Cao and D.D.L. Chung, *Cem. Concr. Res.* 31(11), 1633-1637 (2001).
65. S. Wen and D.D.L. Chung, *J. Electron. Mater.* 30(11), 1448-1451 (2001).
66. D.D.L. Chung, *Smart Mater. Struct.* 4, 59-61 (1995).
67. J. Cao, S. Wen and D.D.L. Chung, *J. Mater. Sci.* 36(18), 4351-4360 (2001).
68. S. Wen and D.D.L. Chung, *Cem. Concr. Res.* 32(3), 335-339 (2002).
69. S. Wen and D.D.L. Chung, *Cem. Concr. Res.*, in press.
70. S. Wen and D.D.L. Chung, *Adv. Cem. Res.*, in press.
71. S. Wen and D.D.L. Chung, *Cem. Concr. Res.* 32(5), 821-823 (2002).
72. S. Wen and D.D.L. Chung, unpublished result.
73. J. Cao and D.D.L. Chung, *Cem. Concr. Res.*, in press.
74. J.-M. Chiou, Q. Zheng and D.D.L. Chung, *Compos.* 20(4), 379-381 (1989).
75. J. Cao and D.D.L. Chung, *Cem. Concr. Res.*, in press.
76. P.-W. Chen and D.D.L. Chung, *Composites: Part B* 27B, 269 (1996).
77. P.-W. Chen and D.D.L. Chung, *ACI Mater. J.* 93(2), 129 (1996).
78. A.M. Brandt and L. Kucharska, *Materials for the New Millennium, Proc. Mater. Eng. Conf.*, ASCE, New York, NY, 1, 271 (1996).
79. N. Banthia and J. Sheng, *Cem. Concr. Compos.* 18(4), 251 (1996).
80. B. Mobasher and C.Y. Li, *ACI Mater. J.* 93(3), 284 (1996).
81. M. Pigeon, M. Azzabi and R. Pleau, *Cem. Concr. Res.* 26(8), 1163 (1996).
82. N. Banthia, C. Yan and K. Sakai, *Cem. Concr. Res.* 20(5), 393 (1998).
83. T. Urano, K. Murakami, Y. Mitsui and H. Sakai, *Compos. – Part A: Applied Science & Manufacturing* 27(3), 183 (1996).
84. P.-W. Chen, X. Fu and D.D.L. Chung, *ACI Mater. J.* 94(3), 203 (1997).

6

Cement-Based Materials for Dielectric Functions

6.1   Dielectric functions

Dielectric materials refer to materials that are poor in electrical conductivity. Their applications include electrical insulation, capacitors (for electrical energy storage), microelectronic substrates, piezoelectricity and pyroelectricity.

Pyroelectricity refers to electricity generated by a change in temperature; it can be used to sense temperature or heat sources (such as people). The direct piezoelectric effect refers to electricity generated by a stress; it can be used to sense a stress or a strain. The reverse piezoelectric effect refers to the generation of strain by an electric field, due to the effect of the electric field on the polarization. Polarization refers to the separation between the positive and negative charge centers of a material. The reverse piezoelectric effect is used for actuation. Sensing and actuation are basic functions of a typical smart structure, which responds to a stress or strain that it senses by providing actuation.

6.2   Dielectric constant

Electrical insulators are also known as dielectrics. Most ionic solids and molecular solids are insulators because of the negligible concentration of conduction electrons or holes.

Consider two metal (conductor) plates connected to the two ends of a battery of voltage V (Fig. 6.1). Assuming that the electrical resistance of the connecting wires is negligible, the potential between the two metal plates is V. Assume that the medium between the plates is a vacuum.

There are a lot of conduction electrons in the metal plates and the metal wires. The positive end of the battery attracts conduction electrons, making the left plate positively charged (charge = +Q). The negative end of the battery repels conduction electrons, making the right plate negatively charged (charge = -Q). Let the area of each plate be A. The magnitude of charge per unit area of each plate is known as the charge density (D_o).

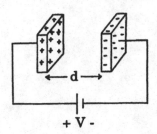

Fig. 6.1   A pair of positively charged and negatively charged conductor plates in vacuum.

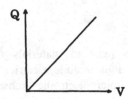

Fig. 6.2   Plot of charge Q vs. potential V. The slope = $C_0 = \dfrac{\varepsilon_o A}{d}$.

$$D_0 = \frac{Q}{A} \ , \tag{6.1}$$

The electric field E between the plates is given by

$$E = \frac{V}{d} \ , \tag{6.2}$$

where d is the separation of the plates.   When E = 0, D_0 = 0.   In fact, D is proportional to E.   Let the proportionality constant be ε_0.   Then

$$D_0 = \varepsilon_0 E \ . \tag{6.3}$$

ε_0 is called the permittivity of free space.   It is a universal constant.

$$\varepsilon_0 = 8.85 \times 10^{-12} \, C/(V.m)$$

Eq. (6.3) is known as Gauss's Law.

Just as D_0 is proportional to E, Q is proportional to V.   The plot of Q versus V is a straight line through the origin (Fig. 6.2), with

$$\text{Slope} = C_0 = \frac{Q}{V} = \frac{\varepsilon_0 EA}{Ed} = \frac{\varepsilon_0 A}{d} \ , \tag{6.4}$$

C_o is known as the capacitance. Its unit is Coulomb/Volt, or Farad (F). In fact, this is the principle behind a parallel-plate capacitor.

Next, consider that the medium between the two plates is not a vacuum, but an insulator whose center of positive charge and center of negative charge coincide when $V = 0$. When $V > 0$, the center of positive charge is shifted toward the negative (right) plate, while the center of negative charge is shifted toward the positive (left) plate. Such displacement of the centers of positive and negative charges is known as polarization. In the case of a molecular solid with polarized molecules (e.g., HF), the polarization in the molecular solid is due to the preferred orientation of each molecule, such that the positive end of the molecule is closer to the negative (right) plate. In the case of an ionic solid, the polarization is due to the slight movement of the cations toward the negative plate and that of the anions toward the positive plate. In the case of an atomic solid, the polarization is due to the skewing of the electron clouds towards the positive plate.

When polarization occurs, the center of positive charge sucks more electrons to the negative plate, causing the charge on the negative plate to be $-\kappa Q$, where $\kappa > 1$ (Fig. 6.3). Similarly, the center of negative charge repels more electrons away from the positive plate, causing the charge on the positive plate to be κQ. κ is a unitless number called the relative dielectric constant. Its value at 1 MHz (10^6 Hz) is 2.3 for polyethylene, 3.2 for polyvinyl chloride, 6.5 for Al_2O_3, 3000 for $BaTiO_3$, and 78.3 for water.

The charges in the plates when a vacuum is between the plates are called free charges (magnitude $= Q$ on each plate). The extra charges in the plates when an insulator is between the plates are called the bound charges (magnitude $= \kappa Q - Q = (\kappa-1)Q$ on each plate).

When an insulator is between the plates, the charge density is given by

$$D_m = \kappa D_o = \frac{\kappa Q}{A} \qquad (6.5)$$

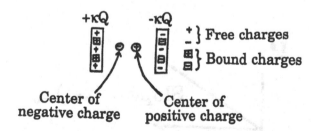

Center of negative charge   Center of positive charge

Fig. 6.3 A pair of positively charged and negatively charged conductor plates in a medium with relative dielectric constant κ.

Using Eq. (6.3), Eq. (6.5) becomes

$$D_m = \kappa \varepsilon_o E = \varepsilon E \ , \tag{6.6}$$

where $\varepsilon \equiv \kappa \varepsilon_o$; ε is known as the dielectric constant, whereas κ is known as the relative dielectric constant. Hence, ε and ε_o have the same unit.

When an insulator is between the plates, the capacitance is given by

$$C_m = \frac{\kappa Q}{V} = \frac{\kappa \varepsilon_o EA}{Ed} = \frac{\kappa \varepsilon_o A}{d} = \kappa C_o \ . \tag{6.7}$$

From Eq. (6.7), the capacitance is inversely proportional to d, so capacitance measurement provides a way to detect changes in d (i.e., to sense strain).

Mathematically, the polarization is defined as the bound charge density, so that it is given by

$$\begin{aligned} P &= D_m - D_o \\ &= \kappa \varepsilon_o E - \varepsilon_o E \\ &= (\kappa - 1) \varepsilon_o E \end{aligned} \tag{6.8}$$

The plot of P versus E is a straight line through the origin, with a slope of $(\kappa-1) \varepsilon_o$ (Fig. 6.4).

The ratio of the bound charge density to the free charge density is given by

$$\frac{\kappa Q - Q}{Q} = \kappa - 1 \ . \tag{6.9}$$

The quantity $\kappa-1$ is known as the electric susceptibility (χ). Hence,

$$\chi = \kappa - 1 = \frac{P}{\varepsilon_o E} \ . \tag{6.10}$$

The dipole moment in the polarized insulator is given by

$$(\text{bound charge})d = (\kappa - 1) Qd \ , \tag{6.11}$$

since the bound charges are induced by the dipole moment in the polarized insulator.

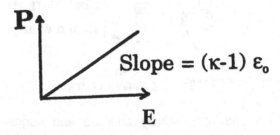

Fig. 6.4   Plot of polarization **P** vs. electric field E.

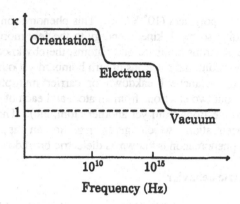

Fig. 6.5 Dependence of the relative dielectric constant κ on the frequency.

The dipole moment per unit volume of the polarized insulator is thus given by

$$\frac{\text{Dipole moment}}{\text{Volume}} = \frac{(\kappa - 1)Qd}{Ad} = \frac{(\kappa - 1)Q}{A} = P \quad . \tag{6.12}$$

Therefore, another meaning of polarization is the dipole moment per unit volume.

Now consider that the applied voltage V is an AC voltage, so that it alternates between positive and negative values. When V > 0 and E > 0, the polarization is one way. When V < 0 and E < 0, the polarization is in the opposite direction. If the frequency of V is beyond about 10^{10} Hz (Hz = cycles per s, s = seconds), the molecules in the insulator cannot reorient themselves fast enough to respond to V (i.e., dipole friction occurs), so κ decreases (Fig. 6.5). When the frequency is beyond 10^{15} Hz, even the electron clouds cannot change their skewing direction fast enough to respond to V, so κ decreases further. The minimum value of κ is 1, which is the value for vacuum.

In making capacitors, one prefers to use insulators with very large values of κ, so that C_m is large. However, one should be aware that the value of κ depends on the frequency. It is challenging to make a capacitor that operates at very high frequencies.

The value of κ also changes with temperature and with stress for a given material.

6.3 Dielectric strength

The maximum electric field that an insulator can withstand before it loses its insulating behavior is known as the dielectric strength. It is lower for ceramics

$(10^4 - 10^7$ V/cm) than polymers $(10^8$ V/cm). This phenomenon is due to the large electric field providing so much kinetic energy to the few mobile electrons in the insulator that these electrons bombard other atoms, thereby knocking electrons out of these atoms. The additional electrons in turn bombard yet other atoms, resulting in a process called avalanche breakdown or carrier multiplication (since one electron may knock out two electrons from an atom and each of these two electrons may knock out two electrons from yet another atom, etc). The consequence is a large carrier concentration, which gives rise to an appreciable electrical conductivity. This phenomenon is known as dielectric breakdown.

6.4   Piezoelectric behavior

A piezoelectric material [1-18] is a material which exhibits one or both of the following effects. In one effect, called the direct piezoelectric effect, the relative dielectric constant κ of a material changes in response to stress or strain; the effect allows the sensing of stress or strain through electrical measurement. In the other effect, called the reverse (or converse) piezoelectric effect, the strain of a material changes in response to an applied electric field; the effect allows actuation that is controlled electrically. Both effects are reversible. Reversibility is valuable for multiple use (not just once) in sensing or actuation.

Both effects are associated with the electric dipole moment in the material. In the direct piezoelectric effect, the dipole moment per unit volume changes in response to strain, due to the movement of ions or functional groups in the material. The change in dipole moment per unit volume affects the capacitance which in turn affects the reactance (imaginary part of the complex impedance $Z = R + jX$, where R is the resistance, X is the reactance, and $j = \sqrt{-1}$). An RLC meter can be used to measure the capacitance from which κ can be obtained using Eq. (6.7).

In the reverse piezoelectric effect, the dipole moment per unit volume changes in response to an electric field and this change is accompanied by the movement of ions or functional groups, thereby causing strain. The change in dipole moment per unit volume may be due to the change in orientation of electric dipoles in the material. It may also be due to the change in the dipole moment for each dipole.

The most common piezoelectric materials are nonstructural materials in the form of ionic crystalline materials whose crystal structure (unit cell) does not have a center of symmetry, so that the centers of positive and negative ionic charges do not coincide and an electric dipole (the magnitude of charge multiplied by the distance between the centers of positive and negative charges) is resulted. This situation occurs either with no applied pressure (in the case of barium titanate) or with an applied pressure (in the case of quartz). The dipole moment per unit volume is the polarization $\mathbf{P}$ (Eq. (6.12)). Every unit cell is associated with a dipole moment. For a column of unit cells in a single crystal, the opposite charges at the adjacent ends of the neighboring unit cells cancel one another, leaving only the charge at the top of the top unit cell and that at the bottom of the bottom unit cell, as illustrated in Fig. 6.6. These charges that are not cancelled are called surface charges, which give rise to the bound charges in each metal plate sandwiching the piezoelectric material.

Under a mechanical force (i.e., an applied stress), the unit cell is slightly distorted so that the dipole moment per unit volume (i.e., the polarization) is changed and the amount of surface charges is changed (Fig. 6.7 (a)). This change in surface charge amount causes a current pulse to flow from one plate to the other through the external circuit which electrically connects the two plates (Fig. 6.7 (a) and (b)). The current pulse can be converted to a voltage pulse by using a resistor in the external circuit. If the stress is varying with time at a sufficiently high rate, the current varies with time exactly like how the stress varies with time (Fig. 6.7 (c)), because time does not allow the observation of the tail of a current pulse. Fig. 6.7 (c) corresponds to a common way in which a piezoelectric sensor is used in practice for sensing dynamic strain. This means that only time-varying stress at a high enough frequency (more than about 0.01 Hz) can be sensed using the method of Fig. 6.7(a). On the other hand, if the two plates are not connected electrically, no current can flow and the increase in polarization due to the applied stress causes extra charge on each plate and hence a voltage across the two plates, as illustrated in Fig. 6.7(d). A static stress causes a static voltage, whereas a time-varying stress causes a time-varying voltage. Thus, the method of Fig. 6.7(d) allows the sensing of both static and dynamic strain. This phenomenon (whether Fig. 6.7(a) or Fig. 6.7(d)) is the direct piezoelectric effect. It allows the conversion (transduction) of mechanical energy to electrical energy for the purpose of sensing. Applications are the detection of mechanical vibrations, force measurement and the detection of audible frequencies.

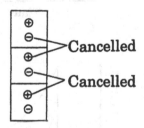

Fig. 6.6   A column for three unit cells in a piezoelectric single crystal. Each unit cell has a center of positive charges + and a center of negative chargers –. These centers do not coincide.

As a consequence of the polarization (situation in which the positive and negative charge centers do not coincide), an electric field is present in a piezoelectric material. When an electric field is further applied to the material, the dipole moment in the material has to change, thus causing the unit cell dimension to change slightly (i.e., causing a strain) (Fig. 6.8). This phenomenon is the reverse piezoelectric effect. It allows the transduction of electrical energy into mechanical energy for the purpose of actuation. Applications are the generation of acoustic or ultrasonic waves (as needed in buzzers and ultrasonic cleaners), micropositioning and vibration-assisted machining.

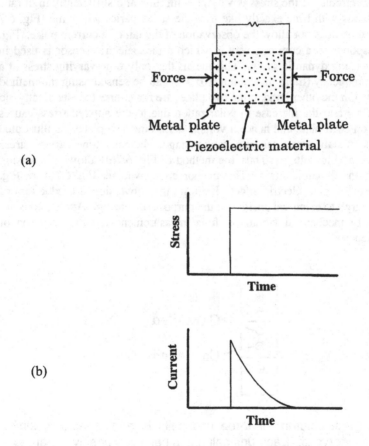

(a)

(b)

Fig. 6.7   (a) The direct piezoelectric effect in which a current pulse results from an applied stress. (b) Current pulse resulting from static stress. (c) Current wave resulting from stress wave. (d) The direct piezoelectric effect in which a voltage results from an applied stress.

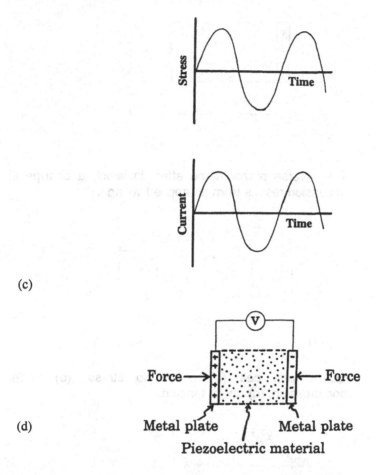

(c)

(d)

Force→ ←Force

Metal plate　　Metal plate
Piezoelectric material

Fig. 6.7 Continued.

An example of a piezoelectric material is quartz (SiO_2), which has Si^{4+} and O^{2-} ions. Without an applied stress, the centers of positive and negative charges overlap (Fig. 6.9(a)). With an applied stress, the centers of positive and negative charges are displaced from one another, such that the displacement is one way under compression (Fig. 6.9(b)) and the other way under tension (Fig. 6.9(c)).

The direct piezoelectric effect of Fig. 6.10 is described by the equation

$$P = d\sigma ,$$ (6.13)

where σ is the applied stress (force per unit area; positive for tensile stress and negative for compressive stress) and **d** is the piezoelectric coupling coefficient (also

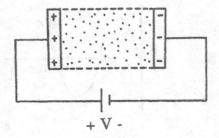

Fig. 6.8   The reverse piezoelectric effect in which a change in dimension results from an applied voltage.

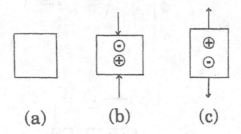

Fig. 6.9   Quartz unit cell (a) under no stress, (b) under compression, (c) under tension.

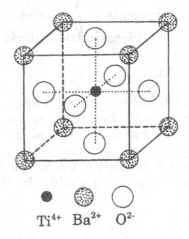

Fig. 6.10   The Perovskite structure.

called the piezoelectric charge coefficient). Eq. (6.13) means that $\mathbf{P}$ is proportional to σ, such that $\mathbf{d}$ is the proportionality constant. From Eq. (6.13), the unit of $\mathbf{d}$ is C/Pa.m^2, since the unit of $\mathbf{P}$ is C/m^2 and the unit of σ is Pa. In general, $\mathbf{P}$ and σ are not necessarily linearly related, so Eq. (6.13) should really be written as

$$\partial \mathbf{P} = \mathbf{d}\,\partial\sigma. \tag{6.14}$$

In the determination of $\mathbf{d}$, the applied electric field Σ is fixed while κ is measured as a function of σ. Hence, from Eq. (6.14) and (6.8),

$$\mathbf{d} = \varepsilon_o \mathrm{E}\left|\frac{\partial\kappa}{\partial\sigma}\right|. \tag{6.15}$$

The direct piezoelectric effect involves a change in stress causing a change in electric field, thereby causing a change in voltage. Knowing $\mathbf{d}$ and κ, one can calculate the change in voltage due to a change in stress $\partial\sigma$, as explained below.

From Eq. (6.8),

$$\partial \mathrm{E} = \frac{\partial \mathbf{P}}{\varepsilon_o(\kappa-1)}. \tag{6.16}$$

Using Eq. (6.14), Eq. (6.16) becomes

$$\partial \mathrm{E} = \frac{\mathbf{d}\partial\sigma}{\varepsilon_o(\kappa-1)}. \tag{6.17}$$

The change in voltage $\partial \mathrm{V}$ is given by

$$\partial \mathrm{V} = \ell\partial \mathrm{E}, \tag{6.18}$$

where ℓ is the length of the specimen in the direction of polarization. From Eq. (6.17) and (6.18),

$$\partial \mathrm{V} = \frac{\ell \mathbf{d}\partial\sigma}{\varepsilon_o(\kappa-1)}. \tag{6.19}$$

The piezoelectric voltage coefficient $\mathbf{g}$ is defined as

$$\mathbf{g} = \frac{\mathbf{d}}{(\kappa-1)\varepsilon_o}. \tag{6.20}$$

From Eq. (6.19) and (6.20),

$$\partial \mathrm{V} = \ell \mathbf{g}\partial\sigma \tag{6.21}$$

Table 6.1   The piezoelectric constant **d** (longitudinal) for selected materials

Material	Piezoelectric constant **d** $(C/N = m/V)$
Quartz	2.3×10^{-12}
$BaTiO_3$	100×10^{-12}
$PbZrTiO_6$	250×10^{-12}
$PbNb_2O_6$	80×10^{-12}

For a large value of ∂V, **g** should be large. From Eq. (6.20), for **g** to be large, **d** should be large and κ should be small. Since **P** and σ do not have to be in the same direction, **d** is a tensor quantity. Consider three orthogonal axes: x_1, x_2 and x_3. When both **P** and σ are in the same direction, **d** is called longitudinal **d**. (When both **P** and σ are along x_3, the longitudinal **d** is called d_{33}. Similar definitions apply to d_{22} and d_{11}.) When **P** and σ are in two orthogonal directions, **d** is called the transverse **d**. (When **P** is along x_3 and σ is along x_1, the transverse **d** is called d_{31}. Similar definitions apply to d_{32}, d_{21}, d_{23}, d_{12}, and d_{13}.) When the applied stress is not tensile or compressive, but shear about a certain direction (i.e., forces perpendicular to a certain direction) and **P** is along a different direction, **d** is called the shear **d**. (When the shear stress is about x_2 and **P** is along x_1, the shear **d** is called d_{15}, in order to distinguish from the transverse d_{12}.) The various **d** values can be positive or negative. Values of the longitudinal **d** for selected materials are shown in Table 6.1.

The reverse piezoelectric effect of Fig. 6.8 is described by the equation

$$S = dE , \qquad (6.22)$$

where **S** is the strain (fractional increase in length). From Eq. (6.22), the unit of **d** is m/V, since the unit of E is V/m and **S** has no unit. Note that V/m is the same as $C/Pa.m^2$, since $\dfrac{C}{Pa.m^2} = \dfrac{C}{N} = \dfrac{C}{J/m} = \dfrac{J/V}{J/m} = \dfrac{m}{V}$. The **d** in Eq. (6.23) is the same as that in Eq. (6.22). These two equations are different ways of expressing the same science. In general, **S** and E are not necessarily linearly related, so Eq. (6.22) should really be written as

$$\partial S = d\partial E. \qquad (6.23)$$

Dividing Eq. (6.22) by Eq. (6.23) gives

$$\frac{E}{\sigma} = \frac{S}{P} . \qquad (6.24)$$

Equivalently, dividing Eq. (6.24) by Eq. (6.14) gives

$$\frac{\partial E}{\partial \sigma} = \frac{\partial S}{\partial P} \ . \tag{6.25}$$

Using Eq. (6.8), Eq. (6.24) becomes

$$\frac{E}{\sigma} = \frac{S}{(\kappa - 1)\varepsilon_0 E} \tag{6.26}$$

and Eq. (6.25) becomes

$$\frac{\partial E}{\partial \sigma} = \frac{\partial S}{(\kappa - 1)\varepsilon_0 \partial E} \ . \tag{6.27}$$

Using Eq. (6.22) and (6.26), we get

$$\frac{E}{\sigma} = \frac{d}{(\kappa - 1)\varepsilon_0} . \tag{6.28}$$

Using Eq. (6.23) and (6.27), we get

$$\frac{\partial E}{\partial \sigma} = \frac{d}{(\kappa - 1)\varepsilon_0} \ . \tag{6.29}$$

Using Eq. (6.20), Eq. (6.28) becomes

$$E = g\sigma \tag{6.30}$$

and Eq. (6.29) becomes

$$\partial E = g\partial\sigma. \tag{6.31}$$

Like Eq. (6.13), Eq. (6.30) is used to describe the piezoelectric effect of Fig. 6.7. Since g depends on d (Eq. (6.20)), g is not an independent parameter.

Stress (σ) and strain (S) are related by the Young's modulus (Y) through Hooke's law, i.e.,

$$\sigma = YS. \tag{6.32}$$

Hence, Eq. (6.30) can be written as

$$E = gYS. \tag{6.33}$$

Rearrangement gives

$$S = \frac{E}{gY} \ . \tag{6.34}$$

Combination of Eq. (6.34) and (6.22) gives

$$d = \frac{1}{gY}$$

or

$$Y = \frac{1}{gd}. \quad\quad (6.35)$$

Eq. (6.35) means that **g** and **d** are simply related through **Y**.

A large value of **d** is desirable for actuators, and a large value of **g** is desirable for sensors. From Eq. (6.20), a low value of κ is favorable for attaining a high value of **g**.

The *electromechanical coupling factor* or *electromechanical coupling coefficient* (**k**) is defined as

$$k^2 = \frac{\text{output mechanical energy}}{\text{input electrical energy}}. \quad\quad (6.36)$$

As the effects are reversible,

$$k^2 = \frac{\text{output electrical energy}}{\text{input mechanical energy}}.$$

The factor **k** describes the efficiency of the energy conversion with a piezoelectric transducer. For a piezoelectric material, **k** is typically less than 0.1. For a ferroelectric material (Sec. 6.4), **k** is typically between 0.4 and 0.7 (0.38 for $BaTiO_3$ and 0.66 for $PbZrO_3$-$PbTiO_3$ solid solution).

The direct piezoelectric effect is useful for strain sensing, which is needed for structural vibration control, traffic monitoring and smart structures in general. The effect is also useful for conversion from mechanical energy to electrical energy, as in converting the stress exerted by a vehicle running on a highway to electricity. This effect is well-known for ceramic materials that exhibit the distorted Perovskite structure. These ceramic materials are expensive and are not rugged mechanically, so they are not usually used as structural materials. Devices made from these ceramic materials are commonly embedded in a structure or attached on a structure. Such structures tend to suffer from high cost and poor durability. The use of the structural material itself (e.g., concrete) to provide the piezoelectric effect is desirable, as the structural material is low-cost and durable.

The direct piezoelectric effect occurs in cement and is particularly strong when the cement contains short steel fibers and polyvinyl alcohol (PVA) [17]. The longitudinal piezoelectric coupling coefficient **d** is 2×10^{-11} m/V and the piezoelectric voltage coefficient **g** is 9×10^{-4} m^2/C. These values are comparable to those of commercial ceramic piezoelectric materials such as lead zirconotitanate PZT (Table 6.2). The effect in cement is attributed to the movement of the mobile ions in cement.

Table 6.2   Measured longitudinal piezoelectric coupling coefficient
d, measured relative dielectric constant κ, calculated piezoelectric
voltage coefficient g and calculated voltage change resulting from a
stress change of 1 kPa for a specimen thickness of 1 cm in the direction
of polarization

Material	d (10^{-13} m/V)*	κ[†]	g (10^{-4} m^2/C)[†]	Voltage change (mV)[†]
Cement paste (plain)	0.659 ± 0.031	35	2.2	2.2
Cement paste with steel fibers and polyvinyl alcohol	208 ± 16	2700	8.7	8.7
Cement paste with carbon fibers	3.62 ± 0.40	49	8.5	8.5
PZT	136	1024	15	15

* Averaged over the first half of the first stress cycle
[†] At 10 kHz

Another method of attaining a structural composite that is itself
piezoelectric is to use piezoelectric fibers as the reinforcement in the composite.
However, fine piezoelectric fibers with acceptable mechanical properties are far
from being well developed. Their high cost is another problem.

Yet another method to attain a structural composite that is itself
piezoelectric is to use conventional reinforcing fibers (e.g., carbon fibers) while
exploiting the polymer matrix for the reverse piezoelectric effect. Due to some
ionic character in the covalent bonds, some polymers are expected to polarize in
response to an electric field, thereby causing strain. The polarization is
enhanced by molecular alignment, which is in turn enhanced by the presence of
the fibers. This method involves widely available fibers and matrix and is
thereby attractive economically and practically. The reverse piezoelectric effect
occurs in the through-thickness direction of a continuous carbon fiber nylon-6
matrix composite [18]. The longitudinal piezoelectric coupling coefficient is 2.2
$\times 10^{-6}$ m/V, as determined up to an electric field of 261 V/m. The coefficient is
high compared to values for conventional ceramic piezoelectric materials (e.g., 1
$\times 10^{-10}$ m/V for BaTiO$_3$). This is due to the difference in mechanism behind the
effect. The high value is attractive for actuation, as it means that a smaller
electric field is needed for the same strain.

Piezoelectric materials in the form of structural composites are in their
infancy of development, but they are highly promising.

6.5   Ferroelectric behavior

A subset of piezoelectric materials is ferroelectric. In other words, a ferroelectric material is also piezoelectric. However, it has the extra ability to have the electric dipoles in adjacent unit cells interact with one another in such a way that adjacent dipoles tend to align themselves. This phenomenon is called self polarization, which results in ferroelectric domains within each of which all the dipoles are in the same direction. As a consequence, even a polycrystalline material can become a single domain.

Quartz is piezoelectric but not ferroelectric. It is a frequently used piezoelectric material due to its availability in large single crystals, its shapeability and its mechanical strength. On the other hand, barium titanate (BaTiO$_3$), lead titanate (PbTiO$_3$) and solid solutions such as lead zirconotitanate (PbZrO$_3$-PbTiO$_3$, abbreviated PZT) and (Pb,La)-(Ti,Zr)O$_3$ (abbreviated PLZT) are both piezoelectric and ferroelectric.

Barium titanate (BaTiO$_3$) exhibits the Perovskite structure (Fig. 6.10) above 120°C and a distorted Perovskite structure below 120°C. Above 120°C, the crystal structure of BaTiO$_3$ is cubic; below 120°C, it is slightly tetragonal (a = b = 3.98 Å, c = 4.03 Å). It is the tetragonal form of BaTiO$_3$ that is ferroelectric, so BaTiO$_3$ is only ferroelectric below 120°C, which is called the Curie temperature. (Jacques and Pierre Curie experimentally confirmed the piezoelectric effect over

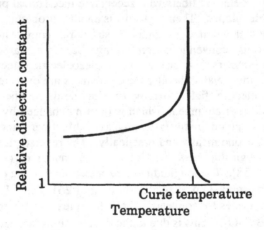

Fig. 6.11   The effect of temperature on the relative dielectric constant of a ferroelectric material. The lowest value of this constant is 1.

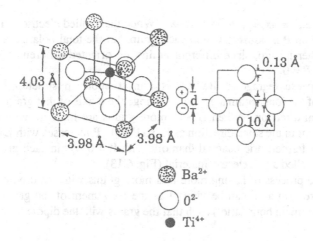

Fig. 6.12 Crystal structure of tetragonal BaTiO$_3$.

100 years ago.) In other words, BaTiO$_3$ exhibits a solid-solid phase transformation at 120°C. As shown in Fig. 6.11, the relative dielectric constant is low above the Curie temperature, high below the Curie temperature and peaks at the Curie temperature (due to the phase transformation). The Curie temperature is 494°C for PbTiO$_3$ and 365°C for PZT (i.e., PbZrO$_3$-PbTiO$_3$ solid solution). It limits the operating temperature range of a piezoelectric material.

In cubic BaTiO$_3$, the centers of positive and negative charges overlap as the ions are symmetrically arranged in the unit cell, so cubic BaTiO$_3$ is not ferroelectric. In tetragonal BaTiO$_3$, the O^{2-} ions are shifted in the negative c-direction, while the Ti^{4+} ions are shifted in the positive c-direction, thus resulting in an electric dipole along the c-axis (Fig. 6.12). The dipole moment per unit cell can be calculated from the displacement and charge of each ion and summing the contributions from the ions in the unit cell. There are one Ba^{2+}, one Ti^{4+} ion and three O^{2-} ions per unit cell (Fig. 6.10). Consider all displacements with respect to the Ba^{2+} ions. The contribution to the dipole moment of a unit cell by each type of ion is listed in Table 6.3. Hence, the dipole moment per unit cell is 17 x 10^{-30} C.m. The polarization **P** is the dipole moment per unit volume, so it is given by the dipole moment per unit cell divided by the volume of a unit cell. Thus,

$$\mathbf{P} = \frac{17 \times 10^{-30}\,\text{C.m}}{4.03 \times 3.98^2 \times 10^{-30}\,\text{m}^3}$$

$$= 0.27\,\text{C.m}^{-2}$$

Consider a polycrystalline piece of tetragonal BaTiO$_3$, such that the grains are oriented with the c-axis of each grain along one of six orthogonal or parallel directions (i.e., +x, -x, +y, -y, +z and –z). When the applied electric field E is zero, the polarization **P** is nonzero within each grain, but the total polarization of all the grains together is zero, since different grains are oriented differently. Therefore, when E = 0, **P** = 0.

Because tetragonal BaTiO$_3$ is almost cubic in structure, very slight movement of the ions within a grain can change the c-axis of that grain to a parallel or orthogonal direction. When E > 0, more grains are lined up with the electric dipole moment in the same direction as E, so **P** > 0. **P** increases with E much more sharply for a ferroelectric material than one with **P** = 0 within each grain at E = 0. The latter is called a paraelectric material (Fig. 6.13).

The process of having more and more grains with the dipole moment in the same direction as E can be viewed as the movement of the grain boundaries (also called domain boundaries) such that the grains with the dipole moment in the

Table 6.3   Contribution to dipole moment of a BaTiO$_3$ unit cell by each type of ion

Ion	Charge (C)	Displacement (m)	Dipole moment (C.m)
Ba^{2+}	$(+2)(1.6 \times 10^{-19})$	0	0
Ti^{4+}	$(+4)(1.6 \times 10^{-19})$	$+0.10(10^{-10})$	6.4×10^{-30}
2O^{2-} (side of cell)	$2(-2)(1.6 \times 10^{-19})$	$-0.10(10^{-10})$	6.4×10^{-30}
O^{2-} (top and bottom of cell)	$(-2)(1.6 \times 10^{-19})$	$-0.13(10^{-10})$	4.2×10^{-30}
			Total = 17×10^{-30}

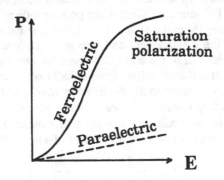

Fig. 6.13   Plots of polarization P vs. electric field E for a ferroelectric material (full line) and a paraelectric material (dashed line).

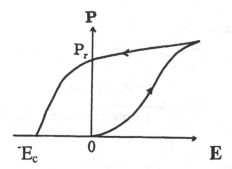

Fig. 6.14 Variation of polarization P with electric field E during increase of E from 0 to a positive value and subsequent decrease of E to a negative value ($-E_c$) at which P returns to 0. Ferroelectric behavior is characterized by a positive value of the remanent polarization and a negative value of $-E_c$.

same direction as E grow while the other grains shrink. These grains are also known as domains (or ferroelectric domains).

When all domains have their dipole moment in the same direction as E, the ferroelectric material becomes a single domain (a single crystal) and **P** has reached its maximum, which is called the saturation polarization. Its value is 0.26 C/m² for $BaTiO_3$ and ~ 0.5 C/m² for PZT.

Upon decreasing E after reaching the saturation polarization, domains with dipole moments not in the same direction as Σ appear again and they grow as E decreases. At the same time, domains with dipole moments in the same direction as E shrink. This process again involves the movement of domain boundaries. In spite of this tendency, **P** does not return all the way to zero when E returns to zero. A remnant polarization ($P = P_r > 0$) remains when E = 0 (Fig. 6.14).

In order for **P** to return all the way to zero, an electric field must be applied in the reverse direction. The required electric field is $E = -E_c$, where E_c is called the coercive field.

When E is even more negative than $-E_c$, the domains start to align in the opposite direction until the polarization reaches saturation in the reverse direction. This is known as polarization reversal. To bring the negative polarization back to zero, a positive electric field is needed. In this way, the cycling of E results in a hysteresis loop in the plot of **P** versus E (Fig. 6.15(a)). The corresponding change in strain **S** associated with the change in polarization is shown in Fig. 6.15(b), which is called the butterfly curve. The slope of the butterfly curve is **d** (Eq. (6.16)).

A ferroelectric material can be used to store binary information, as a positive remnant polarization can represent '0' while a negative remnant

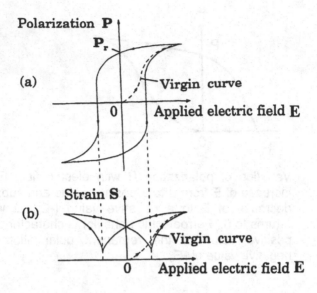

Fig. 6.15   (a) Plot of polarization P vs. electric field E during first application of E (dashed curve) and during subsequent cycling of E (solid curve). (b) Corresponding plot of strain S vs. electric field E.

polarization can represent '1'. The application of an electric field in the appropriate direction can change the stored information from '0' to '1', or vice versa.

Poling refers to the process in which the domains (i.e., dipoles) are aligned so as to achieve saturation polarization. The process typically involves placing the ferroelectric material in a heated oil bath (e.g., 90°C, below the Curie temperature) and applying an electric field. The heating allows the domains to rotate more easily in the electric field. The electric field is maintained until the oil bath is cooled down to room temperature. An excessive electric field is to be avoided as it can cause dielectric breakdown (Sec. 6.3) in the ferroelectric material.

After poling, depoling to a limited extent occurs spontaneously and gradually due to the influence of neighboring domains. This process is known as aging or piezoelectric aging, which has a logarithmic time dependence. The aging rate r is defined as

$$\frac{u_2 - u_1}{u_1} = r \log \frac{t_2}{t_1} , \qquad (6.37)$$

where t_1 and t_2 are the numbers of days after polarization, and u_1 and u_2 are the measured parameters, such as the capacitance.

Through doping (i.e., substituting for some of the barium and titanium ions in $BaTiO_3$ to form a solid solution), the dielectric constant can be increased, the loss factor (related to the difference in phase angle between the electric field and the polarization) can be decreased, and the temperature dependence of the loss factor can be flattened. The Ba^{2+} sites in $BaTiO_3$ are known as A sites; the Ti^{4+} sites are known as B sites. Substitutions for one or both of these sites can occur provided that the overall stoichiometry is maintained.

The most common commercial ferroelectric material is PZT (i.e., $PbZrO_3$-$PbTiO_3$ solid solution or lead zirconotitanate). It is also written as $Pb(Ti_{1-z}Zr_z)O_3$ in order to indicate that a fraction of the B sites is substituted by Zr^{4+} ions. Fig. 6.16 shows the binary phase diagram of the $PbZrO_3$-$PbTiO_3$ system. The cubic phase at high temperatures is paraelectric. The tetragonal and rhombohedral phases at lower temperatures are ferroelectric. The tetragonal phase is titanium rich; the rhombohedral phase is zirconium rich. The tetragonal phase has larger polarization than the rhombohedral phase. A problem with the tetragonal phase is that the tetragonal long axis is 6% greater than the transverse axis, thus causing stress during polarization (so much stress that the ceramic can be shattered). The substitution of Zr for Ti alleviates this problem. The compositions near the morphotropic phase boundary between the rhombohedral and tetragonal phases are those that pole most efficiently. This is because these compositions are associated with a large number of possible poling directions over a wide temperature range. As a result, the piezoelectric coupling coefficient **d** is highest near the morphotropic boundary (Fig. 6.17).

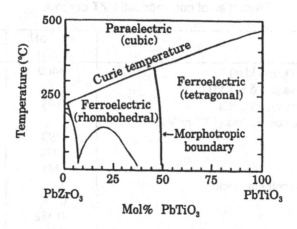

Fig. 6.16   The $PbZrO_3$-$PbTiO_3$ binary phase diagram.

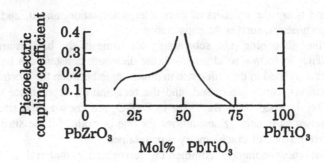

Fig. 6.17   The variation of the piezoelectric coupling coefficient **d** with composition in the $PbZrO_3$-$PbTiO_3$ system.

PZT's are divided into two groups, namely hard PZT's and soft PZT's. This division is according to the difference in piezoelectric properties (not mechanical properties). Hard PZT's generally have a low relative dielectric constant, low loss factor and low piezoelectric coefficients, and are relatively difficult to pole and depole. Soft PZT's generally have a high relative dielectric constant, high loss factor and high piezoelectric coefficients, are relatively easy to pole and depole. Table 6.4 lists the properties of two commercial PZT's, namely PZT-5H (soft PZT) and PZT4 (hard PZT). Hard PZT's are doped with acceptors such as Fe^{3+} for Zr^{4+}, thus resulting in oxygen vacancies. Soft PZT's are doped with donors such as La^{3+} for Pb^{2+} and Nb^{5+} for Zr^{4+}, thus resulting in

Table 6.4   Properties of commericial PZT ceramics

Property	PZT-5H (soft)	PZT4 (hard)
Permittivity (κ at 1 kHz)	3400	1300
Dielectric loss (tan δ at 1 kHz)	0.02	0.004
Curie temperature (T_c, °C)	193	328
Piezoelectric coefficients (10^{-12} m/V)		
$\quad d_{33}$	593	289
$\quad d_{31}$	-274	-123
$\quad d_{15}$	741	496
Piezoelectric coupling factors		
$\quad k_{33}$	0.752	0.70
$\quad k_{31}$	-0.388	-0.334
$\quad k_{15}$	0.675	0.71

A-site vacancies. Hard PZT's typically having a small grain size (about 2 μm), whereas soft PZT's typically have a larger grain size (about 5 μm).

6.6 Effect of admixtures on the dielectric constant of cement-based materials

The static dielectric constant (ε) is a material property that relates to the electric dipole moment per unit volume. It is the product of the permittivity of free space (ε_o) and the relative dielectric constant (κ). The dipole moment per unit volume, also called the polarization, is proportional to κ-1, which is called the electric susceptibility (Eq. (6.10)).

Due to the presence of ionic bonding and moisture in cement, electric dipoles are present and the dielectric constant has been measured for the purpose of fundamental understanding of cement-based materials. Such fundamental studies have addressed the effects of moisture [19-24], chlorides [25], curing age [26-35], aggregate type [25], silica fume [36] and air entrainment [37]. This section is focused on the effect of admixtures, including fibers, latex and silica fume. Carbon fibers are of particular interest in relation to the dielectric effect, due to the ability of carbon fibers to render to cement the ability to sense its own strain by reactance measurement [38]. Reactance is the imaginary part of the complex impedance and is inversely related to the dielectric constant. The effect behind this sensing ability is actually the direct piezoelectric effect [39]. Strain sensing is valuable for smart structures, particularly in relation to structural vibration control.

Dielectric constant measurement using a capacitor probe provides a technique of nondestructive detection of subsurface deterioration [40] and a technique of nondestructive estimation of the compressive strength [21]. Moreover, the dielectric constant is information needed for the use of microwaves (e.g., radars) to inspect concrete structures [41].

Dielectric constant measurement involves subjecting the material under investigation to an AC electric field. One configuration involves the coaxial cable method (i.e., the transmission line method), in which the electromagnetic wave is allowed to propagate and hit the specimen under study [41-43]. The wave can be reflected, absorbed and/or transmitted by the specimen. Another configuration involves that a parallel-plate capacitor [37], in which the AC electric field is applied across the two electrodes of the capacitor while the specimen is sandwiched by the electrodes. The first method is relevant to the use of microwaves to inspect concrete structures. The second method is relevant to the use of a capacitor probe for nondestructive inspection. The second method is used in this section, mainly due to its ability to provide quantitative determination of the dielectric constant.

Table 6.5 [42] shows the relative dielectric constant of seven types of cement paste, each at three frequencies. For any type of cement paste, the relative dielectric constant decreases with increasing frequency, as expected. At any of the frequencies, the addition of silica fume decreases the relative dielectric constant, and the addition of latex increases the relative dielectric constant. The further addition of carbon fibers to a paste containing either silica fume or latex increases the relative dielectric constant. However, an increase in fiber content from 0.5 to 1.0% by mass of cement decreases the relative dielectric contant slightly. All three pastes containing carbon fibers exhibit significantly higher values of the relative dielectric constant than all three pastes which contain no fiber. In contrast, the addition of steel fibers decreases the relative dielectric constant.

The effect of silica fume cannot be due to the air voids, as silica fume is known to decrease the air void content, and air voids tend to decrease the dielectric constant. In other words, a decrease in air void content is expected to increase the dielectric constant. The decrease of the dielectric constant by the silica fume addition is attributed to the low dielectric constant of silica fume (SiO_2) compared to cement; the moisture and ions in cement cause the dielectric constant to be relatively high.

Latex addition increases the dielectric constant of cement paste. This effect is partly attributed to the decrease in air void content (a well-known effect of latex addition). However, the air void content cannot explain the large (up to 27%) increase in the dielectric constant. Latex itself is relatively low in the dielectric constant, since it is a polymer. The observed effect is attributed to the large amount of interface between latex and cement, and the electric dipoles at the interface.

Carbon fiber addition increases the dielectric constant of cement paste. This effect is attributed to the functional groups on the fiber surface (which had been ozone treated) and the resulting dipoles at the fiber-matrix interface. The

Table 6.5   Relative dielectric constant of cement pastes

Paste No.	Fiber type	Fiber content % by mass of cement	vol.%	Admixture	Relative dielectric constant		
					10 kHz	100 kHz	1 MHz
(i)	/	0	0	/	28.6 ± 3.4	24.8 ± 3.6	23.7 ± 2.8
(ii)	/	0	0	SF	20.8 ± 3.4	19.6 ± 3.2	16.5 ± 0.8
(iii)	/	0	0	L	34.9 ± 4.5	31.5 ± 2.9	24.3 ± 2.9
(iv)	Carbon	0.5	0.48	SF	53.7 ± 7.0	38.3 ± 4.8	28.1 ± 2.9
(v)	Carbon	0.5	0.41	L	63.2 ± 5.2	40.4 ± 5.9	33.2 ± 6.8
(vi)	Carbon	1.0	0.95	SF	48.7 ± 4.8	29.6 ± 5.0	25.0 ± 5.0
(vii)	Steel	0.5	0.10	/	19.6 ± 4.8	19.0 ± 1.0	13.7 ± 2.4

SF: silica fume
L: latex

dielectric constant decreases when the fiber volume fraction is increased from 0.48 to 0.95 vol.%, because of percolation. As shown by resistivity data, the percolation threshold is between 0.48 and 0.95 vol.% fibers.

For cement paste containing short steel fibers in the amount of 0.5% by mass of cement (0.10 vol.%, which is much below the percolation threshold) in the absence of silica fume or latex, the relative dielectric constant is lower than that of the paste without any admixture (first entry in Table 6.5), and is comparable to that of the paste with silica fume (second entry in Table 6.5). Hence, steel fiber addition diminishes the relative dielectric constant, in sharp contrast to the effect of carbon fiber addition. The effect of steel fiber addition is attributed mainly to the volume occupied by the fibers in place of cement, although it is also due to the air void content increase (a well-known effect of fiber addition). The contribution of the fiber-matrix interface to the dipole moment is apparently small, due to the oxide on the fiber surface. Hence, the effects on the dielectric constant can be very different for different types of fiber.

Of the seven types of paste listed in Table 6.5, the use of carbon fibers plus latex gives the highest value of the relative dielectric constant, whereas the use of either steel fibers or silica fume gives the lowest value of the relative dielectric constant. A low value is desirable for application as substrates in electronic packaging. A high value is desirable for application as a capacitor for electrical energy storage.

6.7   Effect of stress on the dielectric constant of cement paste

Chapter 3 describes the effect of strain/stress on the electrical resistivity, i.e., the phenomenon known as piezoresistivity, which can be used for strain/stress sensing. In contrast, this section addresses the effect of stress on the dielectric constant [43], i.e., the direct piezoelectric effect [17,44] (Sec. 6.4), which can also be used for stress sensing.

Figures 6.18 and 6.19 show the relative dielectric constant (κ) and the applied stress (negative for compression) during repeated compressive loading of plain cement paste (no admixture) and steel fiber (8 μm diameter) cement paste respectively. For both pastes, κ increases (i.e., the reactance decreases) nonlinearly upon loading. The effect of the compressive strain (value of stress divided by the modulus) on the capacitance is negligible. The greater the stress amplitude, the more κ increases. The longitudinal piezoelectric coupling coefficient d, as averaged over the first half of the first stress cycle, is shown in Table 6.6 for each of four cement pastes. Also shown in Table 6.6 are κ (before stress application), g (calculated by using Eq. (6.20)) and the DC electrical resistivity.

Among the three fibrous admixtures, carbon fibers (1.0 vol.%) results in cement paste with the lowest resistivity, whereas carbon filaments (3 vol.%)

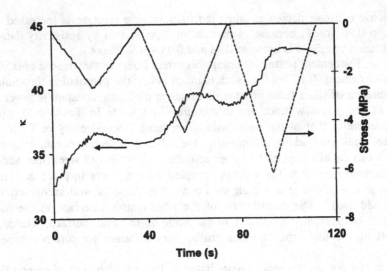

Fig. 6.18 Variation of the relative dielectric constant κ during repeated application of compressive stress for plain cement paste.

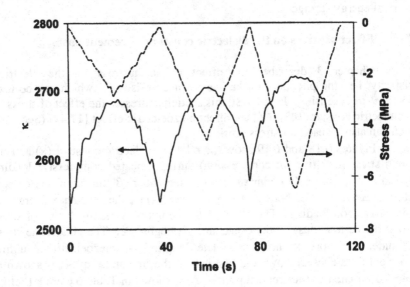

Fig. 6.19 Variation of the relative dielectric constant κ during repeated application of compressive stress for steel fiber (8 μm diameter) cement paste.

Table 6.6   Electrical properties of cement pastes

Cement paste	DC electrical resistivity (Ω.cm)	κ^*	d^* (m/V)	g^* (10^{-3} m^2/C)
Plain cement paste	$(4.8 \pm 0.4) \times 10^5$	33 ± 4	3.1×10^{-13}	1.0
Carbon fiber (15 μm[†]) cement paste	$(8.3 \pm 0.5) \times 10^2$	55 ± 5	4.4×10^{-13}	0.89
Carbon filament (0.1 μm[†]) cement paste	$(1.5 \pm 0.1) \times 10^4$	98 ± 8	4.9×10^{-13}	0.55
Steel fiber (8 μm[†]) cement paste	$(4.5 \pm 0.4) \times 10^3$	2500 ± 200	2.5×10^{-11}	1.1

* At 10 kHz.
† Diameter.

gives the highest resistivity. The high resistivity of the carbon filament cement paste is as described in Ch. 5. It is attributed to the small filament diameter and the consequent large amount of interface between filament and matrix. The interface is associated with a contact resistance. The low resistivity of the carbon fiber cement paste is also as described in Ch. 5. It is attributed to a more optimum fiber diameter. Steel fibers (0.18 vol.%) gives a higher resistivity than carbon fibers (1.0 vol.%), due to the low volume fraction of steel fibers.

The κ value is increased from 33 to 55 by the addition of carbon fibers, but is increased to 90 by the addition of carbon filaments. These increases are attributed to the functional groups at the fiber-matrix interface, which is more abundant in the filament case.

For the steel fiber cement paste, κ reached 2500, in spite of the low volume fraction of steel fibers. A separate experiment indicates that κ is 148 ± 11 at 10 kHz for cement paste containing polyvinyl alcohol (PVA) in the amount of 0.1% by mass of cement, in the absence of steel fibers. Thus, the high κ for the steel fiber cement paste is not due to the PVA, but is probably due to the oxide at the fiber-matrix interface [45].

The increase of κ with stress is quite irreversible for plain cement paste (Fig. 6.18), probably due to the large irreversible nature of the ion movement that occurs during stress application. The behavior is more reversible, but not totally reversible, for cement pastes with carbon fibers, carbon filaments and steel fibers (Fig. 6.19). The greater reversibility in the presence of fibers or filaments is probably due to the reversible nature of the slight movement of functional groups on the fibers or filaments.

The variation of κ with stress is less noisy for the pastes with carbon filaments and steel fibers (Fig. 6.19) than that with carbon fibers. This is probably due to the large interface area provided by the small diameter of the carbon filaments and steel fibers.

The carbon filament cement paste and the steel fiber cement paste are quite effective for stress sensing by κ measurement, though the incomplete reversibility is not desirable and consequently the phenomenon is less effective than piezoresistivity (Ch. 3) for stress sensing. In spite of the good stress-sensing performance, d and g are both low for the carbon filament case.

Fig. 6.20 shows the relationship between κ and the compressive stress for the third stress cycle of the paste with steel fibers. The κ value is lower during loading than during subsequent unloading, as also shown in Fig. 6.19. The third cycle is shown in Fig. 6.20 because it reflects the long-term behavior better than the prior cycles. The relationship between κ and stress is less noisy for the paste with steel fibers than for the paste with carbon filaments. Thus, the effectiveness for stress sensing is better for the paste with steel fibers than that with carbon filaments.

The effect of stress on the dielectric constant is related to the effect of stress on the reactance X [46], which is the imaginary part of the impedance Z (Sec. 6.4).

The effect of stress on the dielectric constant is also related to the effect of stress on the electric polarization [47]. Compressive stress diminishes the extent of electric polarization, as observed in the transverse direction, in cement

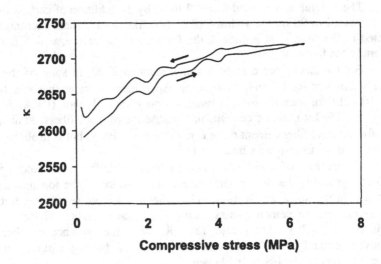

Fig. 6.20   Variation of the relative dielectric constant κ with the compressive stress for the third stress cycle of steel fiber (8 μm diameter) cement paste.

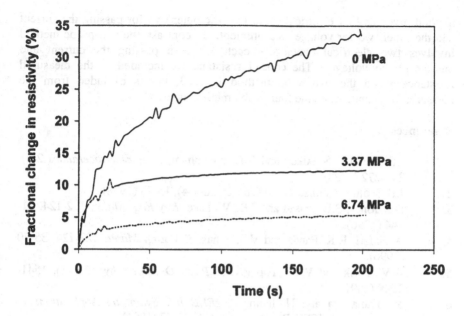

Fig. 6.21   Variation of the measured resistivity with time for plain cement paste.

pastes with and without carbon fibers. In addition, the stress decreases the time for polarization to essentially reach completion. The extent of polarization is much smaller when carbon fibers are present. It is smaller for carbon-fiber cement paste containing silica fume than that containing latex. The time for polarization to reach completion is less than 5 s for carbon fiber silica fume cement paste at a compressive stress of 6.74 MPa. Polarization reversal is hastened by stress.

As an electric field is present during electrical resistivity measurement, polarization can occur in a material during electrical resistivity measurement. As a consequence of the polarization induced electric field in the material being opposite in direction from the applied electric field, polarization causes the measured resistivity to increase with time during resistivity measurement. Hence, polarization can be shown by electrical resistivity measurement conducted over time (Fig. 6.21). Fig. 6.21 also shows that a compressive stress diminishes the extent of polarization.

Polarization can occur within a cement-based material as well as at the interface between the material and an electrical contact. The former dominates when the four-probe method is used for resistivity measurement, whereas the latter dominates when the two-probe method is used [48]. The four-probe method involves four electrical contacts – the outer two for passing the current and the inner two for voltage measurement. In contrast, the two-probe method involves two electrical contacts – each for both passing the current and measuring the voltage. The contact resistance is included in the measured resistance when the two-probe method is used, but is excluded from the measured resistance when the four-probe method is used.

References

1.   J.F. Tressler, S. Alkoy and R.E. Newnham, *J. of Electroceramics* 2(4), 257-272 (1998).
2.   J.F. Scott, *Integrated Ferroelectrics* 20(1-4), 15-23 (1998).
3.   D.J. Jones, S.E. Prasad and J.B. Wallace, *Key Eng. Mater.* 122-124, 71-144 (1996).
4.   A. Safari, R.K. Panda and V.F. Janas, *Key Eng. Mater.* 122-124, 35-70 (1996).
5.   A.V. Turik and V.Yu. Topolov, *J. Phys. D–Appl. Phys.* 30(11), 1541-1549 (1997).
6.   Y. Yamashita and N. Ichinose, *IEEE Int. Symp. on Applications of Ferroelectrics*, IEEE, Piscataway, NJ, 1, 71-77 (1996).
7.   A.V. Turik and V.Yu. Topolov, *Key Eng. Mater.* 132-136(pt 2), 1088-1091 (1997).
8.   V.Ya. Shur and E.L. Rumyantsev, *Ferroelectrics* 191(1-4), 319-333 (1997).
9.   R.G.S. Barsoum, *Smart Mater. & Struct.* 6(1), 117-122 (1997).
10.   C.J. Dias and D.K. Das-Gupta, *IEEE Transactions on Dielectrics & Electrical Insulation* 3(5), 706-734 (1996).
11.   G. Eberle, H. Schmidt and W. Eisenmenger, *IEEE Transactions on Dielectrics & Electrical Insulation* 3(5), 624-646 (1996).
12.   J.F. Scott, *J. Phys. & Chem. of Solids* 57(10), 1439-1443 (1996).
13.   H.B. Harrison, Z-Q. Yao and S. Dimitrijev, *Integrated Ferroelectrics* 9(1-3), 105-113 (1995).
14.   W. Zhong, D. Vanderbilt, R.D. King-Smith and K. Rabe, *Ferroelectrics* 164(1-3), 291-301 (1995).
15.   Q.X. Chen and P.A. Payne, *Measurement Sci. & Tech.* 6(3), 249-267 (1995).
16.   D.J. Jones, S.E. Prasad and J.B. Wallace, *Key Eng. Mater.* 122-124, 71-144 (1996).
17.   S. Wen and D.D.L. Chung, *Cem. Concr. Res.* 32(3), 335-339 (2002).
18.   Z. Mei, V.H. Guerrero, D.P. Kowalik and D.D.L. Chung, *Polym. Compos.* 23(5), 697-701 (2002).
19.   A. Hu, Y. Fang, J.F. Young and Y.-J. Oh, *J. Amer. Ceramic Soc.* 82(7), 1741-1747 (1999).

20. P. Gu and J.J. Beaudoin, *J. Mater. Sci. Lett.* 15(2), 182-184 (1996).

21. R. Zoughi, S.D. Gray and P.S. Nowak, *ACI Mater. J.* 92(1), 64-70 (1995).

22. V. Janoo, C. Korhonen and M. Hovan, *J. Transportation Eng.* 125(3), 245-249 (1999).

23. M. Keddam, H. Takenouti, X.R. Novoa, C. Andrade and C. Alonso, *Cem. Concr. Res.* 27(8), 1191-1201 (1997).

24. C. Alonso, C. Andrade, M. Keddam, X.R. Novoa and H. Takenouti, *Mater. Sci. Forum* 289-292(pt. 1), 15-28 (1998).

25. I.L. Al-Qadi, R.H. Haddad and S.M. Riad, *J. Mater. Civil Eng.* 9(1), 29-34 (1997).

26. A. van Beek and M.A. Hilhorst, *Heron* 44(1), 3-17 (1999).

27. Y. El Hafiane, A. Smith, P. Abelard, J.P. Bonnet and P. Blanchart, *Ceramics-Silikaty* 43(2), 48-51 (1999).

28. S.J. Ford, J.-H. Hwang, J.D. Shane, R.A. Olson, G.M. Moss, H.M. Jennings and T.O. Mason, *Adv. Cem. Based Mater.* 5(2), 41-48 (1997).

29. N. Miura, N. Shinyashiki, S. Yagihara and M. Shiotsubo, *J. Amer. Ceramic Soc.* 81(1), 213-216 (1998).

30. X.Z. Ding, X. Zhang, C.K. Ong, B.T.G. Tan and J. Yang, *J. Mater. Sci.* 31(20), 5339-5345 (1996).

31. X. Zhang, X.Z. Ding, C.K. Ong, B.T.G. Tan and J. Yang, *J. Mater. Sci.* 31(5), 1345-1352 (1996).

32. S.S. Yoon, H.C. Kim and R.M. Hill, *J. Physics D – Applied Physics* 29(3), 869-875 (1996).

33. I.L. Al-Qadi, O.A. Hazim, W. Su and S.M. Riad, *J. Mater. in Civil Eng.* 7(3), 192-198 (1995).

34. X. Zhang, X.Z. Ding, T.H. Lim, C.K. Ong, B.T.G. Tan and J. Yang, *Cem. Concr. Res.* 25(5), 1086-1094 (1995).

35. R.A. Olson, G.M. Moss, B.J. Christensen, J.D. Shane, R.T. Coverdale, E.J. Garboczi, H.M. Jennings and T.O. Mason, *Proc. 1994 MRS Fall Meeting, Microstructure of Cement-Based Systems/Bonding and Interfaces in Cementitious Materials*, Materials Research Society, Warrendale, PA, 370, 255-264 (1995).

36. P. Gu and J.J. Beaudoin, *Adv. Cem. Res.* 9(33), 1-8 (1997).

37. R.H. Haddad and I.L. Al-Qadi, *Proc. 1996 4th Materials Engineering Conf.* Materials for the New Millennium, ASCE, New York, NY, 2, 1139-1149 (1996).

38. X. Fu, E. Ma, D.D.L. Chung and W.A. Anderson, *Cem. Concr. Res.* 27(6), 845-852 (1997).

39. S. Wen and D.D.L. Chung, *Cem. Concr. Res.* 32(3), 335-339 (2002).

40. B.K. Diefenderfer, I.L. Al-Qadi, J.J. Yoho, S.M. Riad and A. Loulizi, *Proc. 1997 MRS Fall Meeting, Nondestructive Characterization of Materials in Aging Systems*, Materials Research Society, Warrendale, PA, 503, 231-236 (1998).

41.   K.J. Bois, A.D. Benally, P.S. Nowak and R. Zoughi, *Proc. 1999 Subsurface Sensors and Applications,* Proc. SPIE – the International Society for Optical Engineering, 3752, 12-18 (1999).

42.   S. Wen and D.D.L. Chung, *Cem. Concr. Res.* 31(4), 673-677 (2001).

43.   S. Wen and D.D.L. Chung, *Cem. Concr. Res.* 32(9), 1429-1433 (2002).

44.   M. Sun, Q. Liu, Z. Li and Y. Hu, *Cem. Concr. Res.* 30(10), 1593-1595 (2000).

45.   X. Fu and D.D.L. Chung, *Compos. Interfaces* 4(4), 197-211 (1997).

46.   X. Fu, E. Ma, D.D.L. Chung and W.A. Anderson, *Cem. Concr. Res.* 27(6), 845-852 (1997).

47.   S. Wen and D.D.L. Chung, *Cem. Concr. Res.* 31(2), 291-295 (2001).

48.   S. Wen and D.D.L. Chung, *Cem. Concr. Res.* 31(2), 141-147 (2001).

7

Cement-Based Materials for Thermal Engineering

7.1 Introduction

Thermal engineering involves temperature sensing, heating, cooling, heat retention, etc. These issues are relevant to concrete structures, as they pertain to energy saving. Since buildings consume much energy, the energy saving that can result from thermal engineering of structures is tremendous.

Thermal engineering of concrete structures is conventionally achieved by the use of nonstructural materials or devices, such as thermometers for temperature sensing, high thermal mass objects for heat retention, embedded heating coils for heating, hot water pipes for heating, etc. Relatively little attention has been given to the use of concrete (a cement-matrix composite) itself for thermal engineering. The use of a structural material for thermal engineering means the elimination or reduction of the need for peripheral non-structural materials for thermal engineering. As structural materials are inexpensive and durable, this results in reduced cost and enhanced durability. Furthermore, the elimination of embedded objects means the avoiding of mechanical property degradation, as embedded objects are like holes in the structure. In addition, the thermal engineering function provided by a structural material is everywhere – not just here and there.

Cement-matrix composites for thermal engineering are addressed in this chapter, with the focus on temperature sensing, resistance (Joule) heating and heat retention. The materials described are structurally superior to conventional cement-matrix composites, while exhibiting the desired functional properties. Moreover, the functional and structural properties do not degrade with time.

7.2 Cement-matrix composites for temperature sensing

The sensing of temperature is needed for thermal control, energy conservation, hazard mitigation and operation control. Concrete structures that benefit from temperature sensing include buildings (for temperature regulation and energy saving), as well as highways, bridges and airport runways (for hazard mitigation and deicing). Temperature sensing is conventionally achieved by thermometers, thermistors or thermocouples in the form of devices that are attached to or embedded in a concrete structure. However, the use of concrete itself for temperature sensing reduces the cost, enhances the durability and allows the sensing function to exist throughout the structure.

Cement-matrix composites for temperature sensing (as thermistors or thermocouples) contain short electrically conductive fibers which are present at a volume fraction below the percolation threshold, so that the fibers essentially do not touch one another. The fibers are not the sensors; they merely modify the composite so that the composite becomes effective for temperature sensing. Although the fiber addition raises the cost of the composite, the cost remains low compared to the use of devices for temperature sensing.

A thermistor is a thermometric device consisting of a material (typically a semiconductor, but in this case a cement paste) whose electrical resistivity decreases with a rise in temperature. A cement-matrix composite containing short carbon fibers is a thermistor due to its resistivity decreasing reversibly with increasing temperature [1] (Sec. 2.5); the sensitivity is comparable to that of semiconductor thermistors.

The Seebeck effect [2-5] is a thermoelectric effect which is the basis for thermocouples for temperature measurement (Sec. 2.8). This effect involves charge carriers moving from a hot point to a cold point within a material, thereby resulting in a voltage difference between the two points.

From the practical point of view, the steel fiber silica fume cement paste containing steel fibers in the amount of 1.0% by weight of cement is particularly attractive for use in temperature sensing, as the absolute thermoelectric power is the highest in magnitude (-68 μV/°C) and the variation of the Seebeck voltage with the temperature difference between the hot and cold ends is reversible and linear. The absolute thermoelectric power is as high as those of commercial thermocouple materials.

By pouring dissimilar cement pastes side by side to make a junction, a cement-based thermocouple with sensitivity 70 μV/°C has been attained [6]. The dissimilar cement pastes are preferably steel fiber cement paste (n-type) and carbon fiber cement paste (p-type).

7.3 Cement-matrix composites for resistance heating

Heating is an important part of the thermal engineering of structures, especially in relation to the heating of buildings and the deicing of bridges and airport runways. Heating can be attained by the flow of hot fluids, the radiation of heat from a space heater, the use of solar energy, etc. It is desirable to use concrete itself for the heating of concrete structures (i.e., self-heating), as this would reduce cost, enhance durability and simplify design. This section describes the use of cement-matrix composites for resistance (Joule) heating.

Resistance heating requires the passage of an electric current. Due to the high electrical resistivity of conventional cement-based materials, the current tends to be low, making resistance heating ineffective. Therefore, cement-matrix composites for resistance heating are those that have relatively low resistivity, as made possible by the use of admixtures, such as short carbon fibers, which are electrically conductive.

Fig. 2.28 [7] gives the volume electrical resistivity of cement-matrix composites containing carbon fibers (~5 mm long, 15 μm diameter) at 7 days of curing. The resistivity decreases much with increasing fiber volume fraction, whether a second filler (silica fume or sand) is present or not.

Please refer to Sec. 5.3 (particularly in connection with Fig. 5.2 and 5.3) for more information on the composition and performance of cement-based materials for resistance heating.

7.4 Cement-matrix composites for heat retention

Heat retention is needed for concrete structures in order to enhance temperature stability. This is particularly important for buildings, for which a high thermal mass is needed. Instead of using nonstructural materials to attain a high thermal mass, it is desirable to use the concrete itself. Since concrete is massively used in a concrete structure, the improvement of the heat retention ability of concrete can be highly effective for increasing the thermal mass of a concrete structure.

Cement-matrix composites for heat retention are those that exhibit relatively high values of the specific heat, as made possible by the use of admixtures. Table 7.1 [8,9] shows the specific heat of cement pastes. The specific heat is significantly increased by the addition of silica fume. It is

Table 7.1 Specific heat (J/g.K, ±0.001) of cement pastes. The value for plain cement paste (with cement and water only) is 0.736 J/g.K

Formulation	As-received silica fume	Silane-treated silica fume
A	0.782	0.788
A$^+$	0.793	0.803
A$^+$F	0.804	0.807
A$^+$O	0.809	0.813
A$^+$K	0.812	0.816
A$^+$S	0.819	0.823

Note:

A: cement + water _ water reducing agent + silica fume
A$^+$:   A + methylcellulose + defoamer
A$^+$F: A$^+$ + as-received fibers
A$^+$O: A$^+$ + O$_3$-treated fibers
A$^+$K: A$^+$ + dichromate-treated fibers
A$^+$S: A$^+$ + silane-treated fibers

Table 7.2   Thermal diffusivity (mm²/s, ±0.03) of cement pastes. The value for plain cement paste (with cement and water only) is 0.36 mm²/s

Formulation	As-received silica fume	Silane-treated silica fume
A	0.26	0.24
A⁺	0.25	0.22
A⁺F	0.27	0.26
A⁺O	0.29	0.27
A⁺K	0.29	0.27
A⁺S	0.25	0.23

Note:

A:	cement + water _ water reducing agent + silica fume
A⁺:	A + methylcellulose + defoamer
A⁺F:	A⁺ + as-received fibers
A⁺O:	A⁺ + O₃-treated fibers
A⁺K:	A⁺ + dichromate-treated fibers
A⁺S:	A⁺ + silane-treated fibers

further increased by the further addition of methylcellulose and defoamer. It is still further increased by the addition of carbon fibers. The effectiveness of the fibers in increasing the specific heat increases in the following order: as-received fibers, O₃-treated fibers, dichromate-treated fibers and silane-treated fibers. This trend applies whether the silica fume is as-received or silane-treated. For any of the formulations, silane-treated silica fume gives higher specific heat than as-received silica fume. The highest specific heat is exhibited by the cement paste with silane-treated silica fume and silane-treated fibers. The specific heat is 12% higher than that of plain cement paste, 5% higher than that of the cement paste with as-received silica fume and as-received fibers, and 0.5% higher than that of the cement paste with as-received silica fume and silane-treated fibers. Hence, silane treatment of fibers is more valuable than that of silica fume for increasing the specific heat.

A low value of the thermal conductivity is also desirable for the purpose of heat retention. The thermal conductivity is the product of the thermal diffusivity, specific heat and density. Table 7.2 [8,9] shows the thermal diffusivity of cement pastes. The thermal diffusivity is significantly decreased by the addition of silica fume. The further addition of methylcellulose and defoamer or the still further addition of fibers has relatively little effect on the thermal diffusivity. Surface treatment of the fibers by ozone or dichromate slightly increases the thermal diffusivity, whereas surface treatment of the fibers by silane slightly decreases the thermal diffusivity. These trends apply whether the silica fume is as-received or silane-treated. For any of the formulations, silane-treated silica fume gives slightly lower (or essentially the same) thermal

Table 7.3 Density (g/cm^3, ± 0.02) of cement pastes. The value for plain cement paste (with cement and water only) is 2.01 g/cm^3

Formulation	As-received silica fume	Silane-treated silica fume
A	1.72	1.73
A$^+$	1.69	1.70
A$^+$F	1.62	1.64
A$^+$O	1.64	1.65
A$^+$K	1.65	1.66
A$^+$S	1.66	1.68

Note:

A:	cement + water _ water reducing agent + silica fume
A$^+$:	A + methylcellulose + defoamer
A$^+$F:	A$^+$ + as-received fibers
A$^+$O:	A$^+$ + O$_3$-treated fibers
A$^+$K:	A$^+$ + dichromate-treated fibers
A$^+$S:	A$^+$ + silane-treated fibers

diffusivity than as-received silica fume. Silane treatments of silica fume and of fibers are about equally effective for lowering the thermal diffusivity.

Table 7.3 [8,9] shows the density of cement pastes. The density is significantly decreased by the addition of silica fume, which is used along with a water reducing agent. It is further decreased slightly by the further addition of methylcellulose and defoamer. It is still further decreased by the still further addition of fibers. The effectiveness of the fibers in decreasing the density decreases in the following order: as-received fibers, O$_3$-treated fibers, dichromate-treated fibers and silane-treated fibers. This trend applies whether the silica fume is as-received or silane-treated. For any of the formulations, silane-treated silica fume gives slightly higher (or essentially the same) specific heat than as-received silica fume. Silane treatment of fibers is more valuable than that of silica fume for increasing the density.

Table 7.4 [8,9] shows the thermal conductivity. It is significantly decreased by the addition of silica fume. The further addition of methylcellulose and defoamer or the still further addition of fibers has little effect on the density. Surface treatment of the fibers by ozone or dichromate slightly increases the thermal conductivity, whereas surface treatment of the fibers by silane has negligible effect. These trends apply whether the silica fume is as-received or silane-treated. For any of the formulations, silane-treated silica fume gives slightly lower (or essentially the same) thermal conductivity as as-received silica fume. Silane treatments of silica fume and of fibers contribute comparably to reducing the thermal conductivity.

Sand is a much more common component in concrete than silica fume. It is different from silica fume in its relatively large particle size and negligible reactivity with cement. Sand gives effects that are opposite from those of silica

Table 7.4   Thermal conductivity (W/m.K, ± 0.03) of cement pastes. The value for plain cement paste (with cement and water only) is 0.53 W/m.K

Formulation	As-received silica fume	Silane-treated silica fume
A	0.35	0.33
A$^+$	0.34	0.30
A$^+$F	0.35	0.34
A$^+$O	0.38	0.36
A$^+$K	0.39	0.37
A$^+$S	0.34	0.32

Note:

A: cement + water _ water reducing agent + silica fume
A$^+$:   A + methylcellulose + defoamer
A$^+$F: A$^+$ + as-received fibers
A$^+$O: A$^+$ + O$_3$-treated fibers
A$^+$K: A$^+$ + dichromate-treated fibers
A$^+$S: A$^+$ + silane-treated fibers

fume, i.e., sand addition decreases the specific heat and increases the thermal conductivity [10].

Table 7.5 [10] shows the thermal behavior of cement pastes and mortars. Comparison of the results on cement paste without silica fume and those on mortar without silica fume shows that sand addition decreases the specific heat by 13% and increases the thermal conductivity by 9%. Comparison of the results on cement paste with silica fume and those on mortar with silica fume shows that sand addition decreases the specific heat by 11% and increases the thermal conductivity by 64%. That sand addition has more effect on the thermal conductivity when silica fume is present than when silica fume is absent is due to the low value of the thermal conductivity of cement paste with silica fume (Table 7.5).

Comparison of the results on cement paste without silica fume and on cement paste with silica fume shows that silica fume addition increases the specific heat by 7% and decreases the thermal conductivity by 38%. Comparison of the results on mortar without silica fume and on mortar with silica fume shows that silica fume addition increases the specific heat by 10% and decreases the thermal conductivity by 6%. Hence, the effects of silica fume addition on mortar and cement paste are in the same direction. That the effect of silica fume on the thermal conductivity is much less for mortar than for cement paste is mainly due to the fact that silica fume addition increases the density of mortar but decreases the density of cement paste (Table 7.5). That the fractional increase in specific heat due to silica fume addition is higher for mortar than cement paste is attributed to the low value of the specific heat of mortar without silica fume (Table 7.5).

Table 7.5   Thermal behavior of cement pastes and mortars

	Cement paste		Mortar	
	Without silica fume*	With silica fume*	Without silica fume*	With silica fume*
Density (g/cm³, ± 0.02)	2.01	1.73	2.04	2.20
Specific heat (J/g.K, ± 0.001)	0.736	0.788	0.642	0.705
Thermal diffusivity (mm²/s, ± 0.03)	0.36	0.24	0.44	0.35
Thermal conductivity§ (W/m.K, ± 0.03)	0.53	0.33	0.58	0.54

* Silane treated
§ Product of density, specific heat and thermal diffusivity

Comparison of the results on cement paste with silica fume and those on mortar without silica fume shows that sand addition gives a lower specific heat than silica fume addition and a higher thermal conductivity than silica fume addition. Since sand has a much larger particle size than silica fume, sand results in much less interface area than silica fume, though the interface may be more diffuse for silica fume than for sand. The low interface area in the sand case is believed to be responsible for the low specific heat and the higher thermal conductivity, as slippage at the interface contributes to the specific heat and the interface acts as a thermal barrier.

Silica fume addition increases the specific heat of cement paste by 7%, whereas sand addition decreases it by 13%. Silica fume addition decreases the thermal conductivity of cement paste by 38%, whereas sand addition increases it by 22%. Hence, silica fume addition and sand addition have opposite effects. The cause is believed to be mainly associated with the low interface area for the sand case and the high interface area for the silica fume case, as explained in the last paragraph. The high reactivity of silica fume compared to sand may contribute to causing the observed difference between silica fume addition and sand addition, though this contribution is believed to be minor, as the reactivity should have tightened up the interface, thus decreasing the specific heat (in contrast to the observed effects). The decrease in specific heat and the increase in thermal conductivity upon sand addition are believed to be due to the higher level of homogeneity within a sand particle than within cement paste.

7.5 Structural performance

The cement-matrix composites mentioned above for thermal engineering are structurally attractive, in addition to being thermally attractive. For example, the improved structural properties rendered by carbon fiber addition pertain to the increased tensile and flexible strengths, the increased tensile ductility and flexural toughness, the enhanced impact resistance, the reduced drying shrinkage and the improved freeze-thaw durability [11-34]. The tensile and flexural strengths decrease with increasing specimen size, such that the size effect becomes larger as the fiber length increases [35]. The low drying shrinkage is valuable for large structures and for use in repair [36,37] and in joining bricks in a brick structure [38,39].

References

1.   S. Wen and D.D.L. Chung, *Cem. Concr. Res.* 29(6), 961-965 (1999).
2.   M. Sun, Z. Li, Q. Mao and D. Shen, *Cem. Concr. Res.* 29(5), 769-771 (1999).
3.   M. Sun, Z. Li, Q. Mao and D. Shen, *Cem. Concr. Res.* 28(12), 1707-1712 (1998).
4.   S. Wen and D.D.L. Chung, *Cem. Concr. Res.* 29(12), 1989-1993 (1999).
5.   S. Wen and D.D.L. Chung, *Cem. Concr. Res.* 30(4), 661-664 (2000).
6.   S. Wen and D.D.L. Chung, *Cem. Concr. Res.* 31(3), 507-510 (2001).
7.   P. Chen and D.D.L. Chung, *J. Electron. Mater.* 24(1), 47-51 (1995).
8.   Y. Xu and D.D.L. Chung, *Cem. Concr. Res.* 29(7), 1117-1121 (1999).
9.   Y. Xu and D.D.L. Chung, *ACI Mater. J.* 97(3), 333-342 (2000).
10.   Y. Xu and D.D.L. Chung, *Cem. Concr. Res.* 30(1), 59-61 (2000).
11.   S. B. Park and B. I. Lee, *Cem. Concr. Compos.* 15(3), 153-163 (1993).
12.   P. Chen, X. Fu and D.D.L. Chung, *ACI Mater. J.* 94(2), 147-155 (1997).
13.   P. Chen and D.D.L. Chung, *Composites* 24(1), 33-52 (1993).
14.   H. A. Toutanji, T. El-Korchi, R. N. Katz and G. L. Leatherman, *Cem. Concr. Res.* 23(3), 618-626 (1993).
15.   N. Banthia, *The Indian Concrete J.* 70(10), 533-542 (1996).
16.   H. A. Toutanji, T. El-Korchi and R. N. Katz, *Cem. Concr. Compos.* 16(1), 15-21 (1994).
17.   S. Akihama, T. Suenaga and T. Banno, *Int. J. Cement Compos. Lightweight Concrete* 8, 21-33 (1986).
18.   M. Kamakura, K. Shirakawa, K. Nakagawa, K. Ohta and S. Kashihara, *Sumitomo kinzoku/Sumitomo Metals* 34(4), 69-78 (1982).
19.   A. Katz and A. Bentur, *Cem. Concr. Res.* 24(2), 214-220 (1994).
20.   Y. Ohama, *Carbon* 27(5), 729-737 (1989).

21. Z. Zheng and D. Feldman, *Prog. Polym. Sci.* (Oxford), 20(2), 185-210 (1995).

22. K. Zayat and Z. Bayasi, *ACI Mater. J.* 93(2), 178-181 (1996).

23. S.V. Itti and J.A. Yalamalli, *Recent Trends in Carbon*, Proc. National Conf., ed. O.P. Bahl, Shipra Publ., Delhi, India, p. 459-468 (1997).

24. N. Banthia, A. Moncef, K. Chokri and J. Sheng, *Can. J. Civil Eng.* 21(6), 999-1011 (1994).

25. B. Mobasher and C.Y. Li, *Infrastructure: New Materials and Methods of Repair*, Proc. Mater. Eng. Conf., ASCE, New York, NY, (804), 551-558 (1994).

26. P. Soroushian, M. Nagi and J. Hsu, *ACI Mater. J.* 89(3), 267-276 (1992).

27. P. Soroushian, *Construction Specifier* 43(12), 102-108 (1990).

28. A.K. Lal, *Batiment Int./Building Research & Practice* 18(3), 153-161 (1990).

29. S.B. Park, B. I. Lee and Y. S. Lim, *Cem. Concr. Res.* 21(4), 589-600 (1991).

30. T.-J. Kim and C.-K. Park, *Cem. Concr. Res.* 28(7), 955-960 (1998).

31. P. Soroushian, Mohamad Nagi and A. Okwuegbu, *ACI Mater. J.* 89(5), 491-494 (1992).

32. V.C. Li and K.H. Obla, *Compos. Eng.* 4(9), 947-964 (1994).

33. N. Banthia, ACI SP-142, *Fiber Reinforced Concrete*, J.I. Daniel and S.P. Shah, Ed., ACI, Detroit, MI, pp. 91-120 (1994).

34. M.A. Mansur, M.S. Chin and T.H. Wee, *J. Mater. In Civil Eng.* 11(1), 21-29 (1999).

35. M.R. Nouri and J. Morshedian, *Iranian J. of Polym. Sci. Tech.* 4(1), 56-63 (1995).

36. S.-S. Lin, *SAMPE J.* 30(5), 39-45 (1994).

37. P. Chen, X. Fu and D.D.L. Chung, *Cem. Concr. Res.* 25(3), 491-496 (1995).

38. M. Zhu and D.D.L. Chung, *Cem. Concr. Res.* 27(12), 1829-1839 (1997).

39. M. Zhu, R. C. Wetherhold and D.D.L. Chung, *Cem. Concr. Res.* 27(3), 437-451 (1997).

8

Cement-Based Materials for Damping

8.1 Introduction to damping

The development of materials for vibration and acoustic damping has been focused on metals and polymers [1]. Most of these materials are functional materials rather than practical structural materials due to their high cost, low stiffness, low strength or poor processability. Thus, this chapter uses practical structural materials, such as concrete, as the starting point in the development of materials for damping. This development involves tailoring through composite engineering and results in reduction of the need for nonstructural damping materials.

Composite materials are widely used for structures due to their strength and stiffness. Damping in structures is commonly provided by viscoelastic non-structural materials. Due to the large volume of structural materials in a structure, the contribution of a structural material to damping can be substantial. The durability and low cost of a structural material add to the attraction of using a structural material to enhance damping. By the use of the interfaces and viscoelasticity provided by appropriate components in a composite material, the damping capacity can be increased with a negligible decrease, if any, of the storage modulus. The attaining of a significant damping capacity while maintaining high strength and stiffness [2-4] is the goal of the structural material tailoring described in this chapter.

Vibrations are undesirable for structures, due to the need for structural stability, position control, durability (particularly durability against fatigue), performance, and noise reduction. Vibration reduction can be attained by increasing the damping capacity (which is expressed by the loss tangent, tan δ, and is a measure of a material's ability to dissipate elastic strain energy during mechanical vibration) and/or increasing the stiffness (which is expressed by the storage modulus). The loss modulus is the product of these two quantities and thus can be considered a figure of merit for the vibration reduction ability.

Damping of a structure can be attained by passive or active methods. Passive methods make use of the inherent ability of certain materials (whether structural or nonstructural materials) to absorb the vibrational energy (for example, through mechanical deformation, as in the case of a viscoelastic material), thereby providing passive energy dissipation. Active methods make use of sensors and actuators to attain vibration sensing and activation to suppress the vibration in real time. The sensors and actuators can be piezoelectric devices [5-10]. This review is focused on materials for passive damping, due to its relatively low cost and ease of implementation.

Materials for vibration damping are mainly metals [11-13] and polymers [14-16], due to their viscoelastic character. Rubber is commonly used as a vibration damping material [17,18]. However, viscoelasticity and molecular movements are not the only mechanisms for damping. Defects such as dislocations, phase boundaries, grain boundaries and various interfaces also contribute to damping [19], since defects may move slightly and surfaces may slip slightly with respect to one another during vibration, thereby dissipating energy. Thus, the microstructure greatly affects the damping capacity of a material. The damping capacity depends not only on the material, but also on the loading frequency, as the viscoelasticity as well as defect response depend on the frequency [20]. Moreover, the damping capacity depends on the temperature [20].

Vibration damping is desirable for most structures. It is commonly attained by attaching to or embedding in the structure a viscoelastic layer [21-25]. However, due to the low strength and modulus of the viscoelastic material compared to the structural material, the presence of the viscoelastic material (especially if it is embedded) lowers the strength and modulus of the structure. A more ideal way to attain vibration damping is to tailor the structural material itself, so that it maintains its high strength and modulus while providing damping.

If an applied stress varies with time in a sinusoidal manner, the sinusoidal stress may be written as

$$\sigma = \sigma_o \sin \omega t \qquad (8.1)$$

where ω is the angular frequency in radians ($\omega = 2\pi v$, where v is the frequency). For Hookian solids, with no energy dissipated, the strain is given by

$$\varepsilon = \varepsilon_o \sin \omega t. \qquad (8.2)$$

The stress and strain are in general not in phase. The strain lags behind the stress by the phase angle δ. The phase angle defines an in-phase and out-of-phase components of the stress, σ' and σ'', i.e.,

$$\sigma' = \sigma_o \cos \delta \qquad (8.3)$$

$$\sigma'' = \sigma_o \sin \delta. \qquad (8.4)$$

The dynamic moduli can be written as

$$E' = \frac{\sigma'}{\varepsilon_o} = E^* \cos\delta \qquad (8.5)$$

$$E'' = \frac{\sigma''}{\varepsilon_o} = E^* \sin\delta. \qquad (8.6)$$

In the complex notation,

$$E^* = E' + jE'' \qquad (8.7)$$

where $j = \sqrt{-1}$ and E' is the storage modulus, which is a measure of the energy stored elastically during deformation. It is approximately equivalent to the elastic modulus determined in creep and relaxation experiments for linear viscoelastic behavior. E'' stands for the loss modulus, which is a measure of the energy converted to heat.

$$\frac{E'}{E''} = \tan\delta \qquad (8.8)$$

The tan δ is the loss tangent which quantifies the damping ability of the material as it is the ratio of the energy lost or dissipated to the energy which is stored by the material.

For this chapter, the reported values for storage modulus, E', is the bending modulus defined by

$$E' = \frac{mL^3}{4bh^3}, \qquad (8.9)$$

where m is the initial slope of the stress-displacement curve and L, b, and h are the length, width, and thickness of the beam respectively.

8.2 Cement-based materials for vibration reduction

The dynamic mechanical properties of concrete have received much less attention than the static mechanical properties, in spite of the fact that dynamic loading conditions are commonly encountered in civil infrastructure systems. The dynamic loading can be due to live loads, sound, wind and earthquakes. The dynamic mechanical properties of concrete can be greatly affected by the admixtures [26-33].

The addition of silica fume (SiO_2 a fine particulate, ~0.1 μm size, preferably surface treated) as an admixture in the cement mix results in a large amount of interface and hence a significant increase in the damping capacity and storage modulus for both cement paste (No. 1-4, Table 8.1) and mortar (No. 14 and 15, Table 8.1) [26,32-34]. The addition of latex (styrene butadiene in the form of a particle dispersion) as an admixture also enhances damping (No. 1 and 13, Table 8.1), due to the viscoelastic nature of latex [27]. However, latex is much more expensive than silica fume.

Table 8.1   Damping capacity (tan δ) and storage modulus of cement-based materials at room temperature, as determined by flexural testing (three-point bending).   Note that cement paste has no sand, whereas mortar has sand.

		tan δ		Storage modulus (GPa)		
		0.2 Hz	1.0 Hz	0.2 Hz	1.0 Hz	Ref.
1.	Cement paste (plain)	0.035	<10⁻⁴	1.9	/	26
2.	Cement paste with untreated silica fume[a]	0.082	0.030	12.7	12.1	33
3.	Cement paste with treated[b] silica fume[a]	0.087	0.032	16.8	16.2	33
4.	Cement paste with untreated silica fume[a] and silane[c]	0.055	/	17.9	/	26
5.	Cement paste with untreated carbon fibers[d] and untreated silica fume[a]	0.089	0.033	13.3	13.8	33
6.	Cement paste with untreated carbon fibers[d] and treated [b] silica fume[a]	0.084	0.034	17.4	17.9	33
7.	Cement paste with treated[b] carbon fibers[d] and untreated silica fume[a]	0.076	0.036	17.2	17.7	33
8.	Cement paste with treated[b] carbon fibers[d] and treated[b] silica fume[a]	0.083	0.033	21	22	33
9.	Cement paste with untreated carbon filaments[d] and treated[b] silica fume[a]	0.089	0.035	10.3	10.9	35
10.	Cement paste with treated[b] carbon filaments[d] and treated[b] silica fume[a]	0.106	0.043	11.3	11.4	35
11.	Cement paste with untreated steel fibers[d] and untreated silica fume[a]	0.051	0.012	12.9	13.2	35
12.	Cement paste with untreated steel fibers[e] and untreated silica fume[a]	0.046	0.011	13.0	13.6	35
13.	Cement paste with latex[f]	0.142	0.112	/	/	27
14.	Mortar (plain)	<10⁻⁴	<10⁻⁴	20	26	32
15.	Mortar with treated[b] silica fume[a]	0.011	0.005	32	33	34
16.	Mortar with untreated steel rebars	0.027	0.007	44	44	34
17.	Mortar with sand blasted steel rebars	0.037	0.012	46	49	34
18.	Mortar with untreated steel rebars and treated[b] silica fume	0.027	0.012	47	48	37

[a] 15% by mass of cement
[b] Treated by silane coating
[c] 0.2% by mass of cement
[d] 0.5 vol.%
[e] 1.0 vol.%
[f] 30% by mass of cement

The addition of either sand (No. 1 and 14, Table 8.1) or 0.5 vol.% of a fibrous admixture in the form of 15 μm-diameter untreated carbon fibers (about 5 mm long), 0.1 μm-diameter carbon filaments (> 100 μm long) or 8 μm-diameter steel fibers (6 mm long) (No. 2,3,5-12, Table 8.1) to the cement mix does not help the damping [30-33,35], due to the relatively high damping associated with the inhomogeneity within cement paste. However, the addition of 15 μm-diameter silane-treated carbon filaments enhances the damping slightly, presumably due to the large interface area between the filaments and the cement matrix and the increased contribution of the interface to damping by the silane present at the interface [35]. In spite of the ductility of steel compared to carbon, the steel fibers reduce the damping capacity [35]. Thus, the interface area appears to be more important than the fiber ductility in enhancing damping. The addition of sand greatly reduces the damping capacity, due to the large proportion of sand compared to that of fibers or filaments. However, sand is inexpensive and is needed to diminish the drying shrinkage, and carbon fibers (particularly surface treated and used along with treated silica fume, which helps the fiber dispersion) are useful for increasing the storage modulus (No. 2,3,5-8, Table 8.1), decreasing the drying shrinkage, increasing the flexural strength and toughness and rendering self-sensing ability [36].

Concrete is commonly reinforced with steel reinforcing bars (rebars in short). Both the loss tangent and the storage modulus of mortar are greatly increased by the use of steel rebar (preferably surface treated by sand blasting) in mortar [34] (No. 14,16 and 17, Table 8.1). The addition of silica fume to steel rebar mortar further enhances the storage modulus (No. 16 and 18, Table 8.1) [37].

Comparison of No. 15 and 18 in Table 8.1 shows that the steel rebars enhance both loss tangent and storage modulus of silica fume mortar. However, comparison of No. 2, 11 and 12 shows that the addition of 8 μm-diameter steel fibers to silica fume cement decreases the loss tangent and has almost no effect on the storage modulus. Thus, steel rebars are much more useful than steel fibers for damping and stiffening. This may be due to the presence of surface deformation patterns on the steel rebar and the smoothness of the steel fiber surface. The importance of surface roughness is supported by the effect of sand blasting the steel rebars. The sand blasting enhances both loss tangent and storage modulus, as shown by comparing No. 16 and 17 [34].

Silane coating [33] is an effective surface treatment, due to the hydrophilicity enhanced by silane and the fact that the cement mix is water-based. The enhanced wettability results in better dispersion in the cement mix. The treatment applied to silica fume enhances the loss tangent and storage modulus. When it is applied to carbon fibers, the storage modulus is increased, with negligible effect on the loss tangent.

Silane can be introduced to silica fume cement in two ways: (i) as a coating on the silica fume (i.e., coating the silica fume with silane prior to using the silica fume (No. 3, Table 8.1) and (ii) as an admixture (i.e., adding the silane directly into the cement mix) (No. 4, Table 8.1). Both methods enhance the workability of silica fume mortar similarly and increase the tensile and compressive strengths of silica fume cement paste similarly. However, the latter method gives silica fume cement paste of lower compressive ductility, lower damping capacity (No. 3 and 4, Table 8.1), more drying shrinkage, lower air void content, higher density, higher specific heat and greater thermal conductivity [26]. These differences are mainly due to the network of covalent coupling among the silica fume particles in the latter case.

References

1.   D.D.L. Chung, *J. Mater. Sci.* 36(24), 5733 (2001).
2.   Z. Maekawa, H. Hiroyuki and A. Goto, *Nippon Kikai Gakkai Ronbunshu, C Hen.* 60(571), 831 (1994).
3.   I.C. Finegan and R.F. Gibson, *Compos. Struct.* 44(2), 89 (1999).
4.   M.R. Maheri and R.D. Adams, *JSME Int. J. Series A – Solid Mechanics & Mater. Eng.* 42(3), 307 (1999).
5.   J.J. Hollkamp and R.W. Gordon, *Smart Mater. & Struct.* 5(5), 715 (1996).
6.   Y.K. Kang, H.C. Park, J. Kim and S.-B. Seung, *Materials & Design* 23(3), 277 (2002).
7.   S. Raja, G. Prathap and P.K. Sinha, *Smart Mater. & Struct.* 11(1), 63 (2002).
8.   M. Hori, T. Aoki, Y. Ohira and S. Yano, *Composites – Part A: Appl. Sci. & Manufacturing* 32(2), 287 (2001).
9.   M. Arafa and A. Baz, *Compos. Sci. & Tech.* 60(15), 2759 (2000).
10.   M. Arafa and A. Baz, *Composites – PartB: Eng.* 31(4), 255 (2000).
11.   I.G. Ritchie and Z.L. Pan, *Metallurgical Transactions A – Physical Metallurgy & Mater. Sci.* 22A(3), 607 (1991).
12.   K. Fujisawa, M. Taniuchi, T. Kogishi, Y. Sasaki, A. Kobayashi and K. Iwai, *R & D* 44(3), 12 (1994).
13.   J.N. Wei, H.F. Cheng, Y.F. Zhang, F.S. Han, Z.C. Zhou and J.P. Shui, *Mater. Sci. & Eng.* 325(1-2), 444 (2002).
14.   S. Thomas and A. George, *European Polym. J.* 28(11), 1451 (1992).
15.   D.A. Greenhill and D.J. Hourston, *Proc. ACS Division of Polymeric Materials Science and Engineering,* ACS, Books & Journals Division, Washington, DC, 60, 644 (1989).
16.   S. Nazarenko, D. Haderski, A. Hiltner and E. Baer, *Polym. Eng. & Sci.* 35(21), 1682 (1995).
17.   S.N. Ganeriwala, *Proc. SPIE – The International Society for Optical Engineering,* Smart Structures and Materials 1995: Passive Damping,

Society of Photo-Optical Instrumentation Engineers, Bellingham, 1995, 2445, 200 (1995).

18. J.N. Das, N.G. Nair and N. Subramanian, *Plastics Rubber & Composites Processing & Applications* 20(4), 249 (1993).
19. R. Schaller and G. Fantozzi, *Mater. Sci. Forum* 366-368, 615 (2001).
20. R. de Batist, *J. de Physique* (Paris), Colloque C9, 44(12), 39 (1983).
21. M.D. Rao, R. Echempati and S. Nadella, *Composites – Part B: Eng.* 28(5-6), 547 (1997).
22. J.M. Biggerstaff and J.B. Kosmatka, *J. Compos. Mater.* 33(15), 1457 (1999).
23. R.F. Gibson and H. Zhao, *American Society of Mechanical Engineers, Noise Control & Acoustics Division* (Publication), 26, 333 (1999).
24. J. Oborn, H. Bertilsson and M. Rigdahl, *J. Appl. Polym. Sci.* 80(14), 2865 (2001).
25. N. Nugay and B. Erman, *J. Appl. Polym. Sci.* 79(2), 366 (2001).
26. Y. Xu and D.D.L. Chung, *Cem. Concr. Res.* 30(8), 1305 (2000).
27. X. Fu, X. Li and D.D.L. Chung, *J. Mater. Sci.* 33, 3601 (1998).
28. N. Moiseev, *Proc. SPIE – the International Society for Optical Engineering,* Vibration Control in Microelectronics, Optics, and Metrology, International Society for Optical Engineering, Bellingham, WA, 1619, 192-202 (1992).
29. X. Fu and D.D.L. Chung, *Cem. Concr. Res.* 26, 69 (1994).
30. X. Fu and D.D.L. Chung, *ACI Mater. J.* 96(4), 455 (1999).
31. Y. Xu and D.D.L. Chung, *Cem. Concr. Res.* 29(7), 1107 (1999).
32. Y. Wang and D.D.L. Chung, *Cem. Concr. Res.* 28(10), 1353 (1998).
33. Y. Xu and D.D.L. Chung, *ACI Mater. J.* 97(3), 333 (2000).
34. S. Wen and D.D.L. Chung, *Cem. Concr. Res.* 30(2), 327 (2000).
35. J. Cao and D.D.L. Chung, unpublished result.
36. D.D.L. Chung, *Composites: Part B* 31(6-7), 511 (2000).
37. S. Wen and D.D.L. Chung, unpublished result.

Index

Printed in the United States
by Baker & Taylor Publisher Services